Dr. Ashley M. Berge, PhD
AM8 International
QLD, AUSTRALIA
+61 401 814 324
AM8@topicthread.com

The Elite Research Method

A methodological framework applied in the real-world

By
Dr. Ashley M. Berge, PhD

First published in 2018 by Dr. Ashley M. Berge, PhD
www.AM8International.com
Copyright © 2022 Ashley M. Berge
2nd Edition
ISBN: 978-0-9945694-6-2

The moral rights of the author have been asserted.

This book is copyright. Apart from any fair dealing for the purposes of private study, research, criticism or review permitted under the *Copyright Act 1968*, no part may be stored or reproduced by any process without prior written permission. Enquiries should be made to the publisher.

Dedication

This one goes to the athletes. If it wasn't for all the players (athletes) that pursue elite careers, this scientific analysis wouldn't be possible. As a scientist, your elite skill acquisition pushes us further. As a coach, your determination and passion inspire us. Yet as a *former* athlete, you have my utmost admiration.

My dear guide throughout the arduous research process, Dr. Sue Whatman – a supervisor for the ages and mentor throughout, your relentless commitment to this undertaking will always be cherished.

To the theorists who have come before me – if it wasn't for your tireless contribution to the sciences, respectively, this methodological framework wouldn't be complete. Your contribution has allowed our fields to persevere, expand, and ultimately has contributed to the development of *The Elite Research Method: a methodological framework applied in the real-world.*

Science is organised knowledge. Wisdom is organised life.
                — *Immanuel Kant*

Foreword

A moral compass is by far the most endearing, rewarding, yet explosive path to set out on conquering. When looking back, I can see why a portion of the population has no interest in a moral compass, arguably, rightly so. The decision to pursue a moral compass, post-formation, is an internal decision in itself – yet once it has been made, the off switch does not seem to be in reach. Through this inability to switch 'off' has come incredible rewards – right here.

Without my moral compass, it would be unfair to suggest, or let alone report, on the scientific nature of derived complexities in research itself – more so, pull these complex-pieces a part to reshuffle – and place in an order the majority can comprehend. You see, science isn't meant to be complicated, rather its very nature is underpinned by the process, or methodologies. Method is synonymous with methodological, outlining a road map of prior and what is being drawn on to investigate – a thorough systematised approach to ensure what is 'found' is done so in a highly rigorous way. In other words – no cheating. The methodology – the approach that brings theory to life in the real-world through its very application and how we create *new* scientific findings.

This anti-cheating and real-world use form the stem of *The Elite Research Method: a methodological framework applied in the real-world*. But, before it's further introduced, let me share what is synonymous with my writing – a recap of how I got to where I am right now, in its briefest form. Rather, if I may, the underpinnings that have tightened my moral compass along its way to the very publishing of *The Elite Research Method: a methodological framework applied in the real-world* in your very hands.

How I came to be, *recap*:

At 18, I became a Personal Trainer / Sports & Strength Conditioning Coach. I was also a full-time athlete, training for 8 hours a day.

At 19, I was a sponsored Elite Athlete. I also commenced University and maintained my full-time training.

At 20, I became a Level 1 Coach, then went on to become one of Australia's youngest ever Level 2 Coaches permitted to go back-to-back.

Just before I turned 21, I received a scholarship to play and train at one of America's leading Academies, in Texas.

At 21, I was coaching some of America's top national players as a direct result of succumbing to another injury and whilst also playing and training when healthy. That same year I received a full-scholarship to a National Collegiate Athletics Association (NCAA) DII private University that marked against the Harvard Grading Scale.

At 22, I spent part of the summer touring through Europe with American national players. At this time I had also started my Undergraduate degree and had received a Grade Point Average (GPA) of a 4.0 (the highest attainable) in both first and second semesters. This came with multiple Dean's awards.

At 23, I again spent part of the summer touring through Europe with American national players. Due to injury, I had to fly back to Australia and leave my scholarship behind. I spent the majority of the year on crutches unable to walk. This was the year I completed my Undergraduate degree and started to run. This was also the year I had to choose between playing and coaching—I chose coaching.

At 24, I commenced my Masters. I spent the summer, again, touring through Europe with American national players. On return I was asked to be the Director of a tennis program and accepted, being one of the youngest coaches, nationally, to take on such a position. After running for less than a year I managed to clock up 50 km days (not weeks). Just before I turned 25, I managed my biggest week yet: 180 km's in total.

At 25, I registered my business with my Dad and opted not to travel but to keep on top of my Masters (after receiving a GPA of 7.0: the highest attainable). Shortly after, my Dad passed away—just shy of my graduation; four months later, my Grandpa passed away.

At 26, I competed in my first Ultra race (100 km) but had a DNF (Did Not Finish) due to an ongoing hamstring injury. In the same year, I became recognised as a Level 3 Coach (the highest attainable globally). Soon after, I commenced my PhD—I still had questions to answer that my studies until then were unable to address.

At 27, my PhD was under full-swing and I was on track to complete my PhD in the minimum time frame. The same year I was able to sponsor my first athlete.

A fortnight after turning 28, I became unwell. Ultimately, I couldn't run, walk, read, write, do simple tasks, let alone finish the home-stretch of my PhD without near-collapsing. Without going into detail, somehow, I did manage to finish my PhD and in the minimum time frame; with only four months leave taken.

At 29, I had PhD corrections to look after, although the entire year was all about getting my health back. That same year, I competed in my first Ultra since becoming unwell—and I finished all 80 km's of it, even though I tore my tendon at the 20-25 km mark. It was about finishing and I did. Mid-year, I had surgery that could have left me far-worse off (from my ongoing health battle) and I am incredibly fortunate to still be here. Those who were there for me find it hard to fathom that I am still alive—something I still find hard to swallow. My gamble on the surgery paid off. With 15 specialists with varied opinions; the risk paid off. Evidently, following a visit to a dentist, a cranial nerve was impinged in my face (jaw) causing chronic pain (severe 24/7 neuralgia) plus a variety of unpleasant side effects. I can't eat properly due to my gut being disrupted from all the medicines, but, today, I am a heck of a lot better.

At 30, I had already published the first instalment in the Optimal Secrets Series, *The Secrets to Optimal Performance Success*, and later that year, *The Secrets to Optimal Wellbeing* was published. And, almost two years since submission, I was officially awarded my doctorate. It was also a year of tightening the vision for what the science inside my PhD was, and eventually over time, going to bring the world. As a result, *The Science of Elite Performance* was published just shy of turning 31.

At 31 it was all about bringing this vision into fruition – and so the journey began of building a social interface that brought the world together, using 'social' as the vehicle to unify individuals across the world. And with that, 'science' was at its core – using the science model inside my PhD to share not only best practice, but to allow users from around the world to be able to, too. This saw Topicthread (topicthread.com) born. But there was a long road ahead.

At 32, *The Secrets to Optimal Coaching Success* was published early in the year. Since then, 2018 has been about Topicthread: *the world's 1st social compilation with the infrastructure to disrupt the future*. There has been remodelling, redesign and deep thought. Opposed to a simple safe and secure social interface, health and education became core initiatives. A community hub for every person across the globe, progressing towards the delivery of these two overarching fields, housed inside Topicthread. Alas! *The Elite Research Method: a methodological framework applied in the real-world* was sitting idle and needed my attention – which brings us to now. It'll be the second publication of the year by AM8 International, and just when things look to be slowing down, they speed right back up.

The Elite Research Method: a methodological framework applied in the real-world is bringing back the emphasis on science and *how it comes to be*. Despite a clear passion in delivering on real-world applications for science, and simplifying specific principles of performance for ease, the journey continues in delivering what was found – scientifically, and philosophically, derived through the identified approaches. In this, is a personal responsibility to share, and to be shared, to allow those seeking answers, to follow a road-map – however entangled, intentionally, despite its genuine intent.

In order for us to find the answers, we first need to find those missing pieces to the puzzle. Over the years I've personally found this has become a niche – gluing these pieces together to present what has been found, and articulate to allow the broader audience to see clearly enough to comprehend, and do the same – without hiding. The Elite  *Research Method: a methodological framework applied in the real-world, is another pursuit, although vastly different in its content. Methodological in its nature, whilst breaking some rules I believe can, or rather should, be broken when the page, and science, demands it. Despite it all, remember it's not about retention or recollection, but rather a journey towards finding that missing piece, that I near-guarantee in some way, shape or form, lies within The Elite Research Method: a methodological framework applied in the real-world – no matter your pretext.*

Contents

Dedication	v
Foreword	ix
Contents	xiv
List of Appendices	16
List of Figures	17
Glossary	18
Map Introduction	25
Map Briefing: An Overview	30
Research Introduction	31
Considerations for designing research	32
The nature of inquiry informing the research design	33
The research questions	36
Map Briefing: An Overview	38
Data collection methods	40
Interviews	40
Questioning methods: the interview process	41
Initial, structured coach interview questions	44
Adapted, semi-structured coach interview questions	45
Document collection	46
Notational analysis and typical reliability	47
Observational analysis techniques	49
Survey development and explanation	52
Map Briefing: An Overview	55
Observational tool and criteria	58

Groundstroke selection criteria for observational analysis	59
The exclusion of opposition quality in the selection criteria	63
Mapping the data collection tools against the theoretical framework	64
Map Briefing: An Overview	68
Research action steps	71
Ethical clearance and ethical process	73
Data collection process	74
The recruitment process	76
Map Briefing: An Overview	80
Recruiting participants: my reflective journal	83
The scene: where I collected the data	83
Notational analysis procedure	85
Distributing the survey	86
The follow-up: waiting to hear back from participants	91
Map Briefing: An Overview	95
Data analysis steps	97
Notational analysis	97
Survey and interview analysis	98
Research Conclusion	99
Map Conclusion	101
What Now?	107
Books by Dr. B	109
Bibliography	163

List of Appendices

Appendix A Survey – 110

Appendix B Ethical Clearance – 126

Appendix C Informed Consent Form and Information Sheet – 130

Appendix D Email Correspondence – 135

Appendix E Work Log – 157

List of Figures

Figure 1   Baseline Boundaries for capturing the forehand groundstroke in action – depicted as a metre guideline – 60

Figure 2   Baseline Boundaries for capturing the forehand groundstroke in action – depicted as a full court with the player's starting positions (serve/return) – 60

Figure 3   The incorporation of the framework through the data collection process: the five methods of data collection and the type of data that was generated through these methods – 66

Figure 4   Research action plan illustrative of the steps taken during the data collection and research process – 72

Glossary

Abbreviation/term	In full
ATP	Association of Tennis Professionals: the Men's international tennis tour. The best male tennis players in the world compete on the ATP tour.
Balance	"The ability to maintain the centre of body mass over a base of support [and]… is the underlying component of all motor skills," (Verstegen & Marcello, 2001, p.145).
Coaching pedagogy	Derived from the individual's coaching philosophy, shaped by their morals and principles which in turn inform how they go about coaching players/athletes (both teaching and learning).
COG	Centre of Gravity.
Contact above hip height	When the player makes contact with their forehand groundstroke above their hip region (used as a visual point of reference).
Contact at/around hip height	When the player makes contact with their forehand groundstroke at/around hip height, also referred to as centre of gravity (COG) (used as a visual point of reference point), denoted in images at times by a green line.

Abbreviation/term	In full
Contact below hip height	When the player makes contact with their forehand groundstroke below their hips (used as a visual point of reference).
Deep knee bend	Denoted by knee bend and the angle behind the leading leg, corresponding with the racquet arm, with an angle <120 degrees, according to Schonborn's (2000) findings.
Dynamic balance	"Being able to maintain body mass over the base of support while the body is in motion," (Verstegen & Marcello, 2001, p. 145).
Forehand groundstroke	The groundstroke hit with the player's leading/dominant hand, predominantly struck with one hand and contacted after the ball has bounced once.
Grand Slam	The pinnacle of tennis accolades with four major events held annually, comprising the Australian Open, French Open (or Roland Garros), Wimbledon, and the US Open (or Flushing Meadows).
Grand Slam champions	Players who have won a Grand Slam, or coaches who have worked with or currently work with a player who has won a Grand Slam.

Abbreviation/term	In full
Higher ranked players	Players who are ranked closer to 1 with a ranking <50 on either the WTA or ATP tour.
Hip height	A visual reference point for height of point of contact in the forehand groundstroke, interchangeably described as "at" hip height, "around" hip height, or as "at/around" hip height.
ITF	International Tennis Federation.
KP	"Category of augmented feedback that gives information about the movement characteristics that led to a performance outcome," (Magill, 2007, p. 335).
Lower body positioning	The position the player acquires when contacting a groundstroke, specifically referencing their lower body and the knee flexion present.

Abbreviation/term	In full
Lower ranked players	Players who are ranked closer to 100 or higher on either the WTA or ATP tours, also referred to as not being 'as' successful as higher ranked players.
Match	According to the ITF, a tennis match "can be played to the best of 3 sets (a player/team needs to win 2 sets to win the match) or to the best of 5 sets (a player/team needs to win 3 sets to win the match)," (ITF, 2011, p. 6).
Optimal	Used to indicate "best or most favourable" results, performance or coaching practices leading to the highest chance of success (Oxford Dictionary, 2015). Success in this context translates into a higher ranking.
Point of contact	The moment a tennis player's racket face/strings comes into contact with the ball during the forehand groundstroke technique.
Snowball Sampling	"Snowball sampling [is] a method that has been widely used in qualitative research [where it] yields a study sample through referrals made among people who share or know of others who possess some characteristics that are of research interest," (Biernacki & Waldorf, 1981, p. 141).

Abbreviation/term	In full
Slight knee bend	Denoted by knee bend and the angle behind the leading leg, corresponding with the racquet arm, with an angle >121 degrees.
SPSS	Statistical Package for the Social Sciences.
Skill	"An activity or task that has a specific purpose of goal to achieve; an indicator of performance," (Magill, 2007, p. 5).
V Position	The application of 1) knee bend, *and* 2) height of point of contact "around" (interchanged with "at") or "below" the tennis player's hip height when the player comes into contact with the ball, executing their forehand groundstroke with the presence of dynamic balance. The "V" denotes the position of the body behind the knees, at the ankles, and at the player's arm when reaching out to hit the forehand groundstroke (creating a V shape with the leading upper leg muscle/quadricep and point of contact) (Berge, 2014).
WTA	Women's Tennis Association: the Women's international tennis tour. The best female tennis players in the world compete on the WTA tour.

The Elite Research Method

A methodological framework applied in the real-world

**By
Dr. Ashley M. Berge, PhD**

Map Introduction

Seldom does anything work by itself. The age-old argument of *two is better than one* is applicable to the sciences. Often confused, the sciences comprise of a multitude of fields, with each discipline its own respective niche. One discipline without consideration for the next, limits the scope and thereby extension of its data into the *real-world*.

Arguably, not all science is intended to be transitional to the real-world, or to have real-world applications. Those that do have merit for transition, offer a use-case for application in *real-world*. This type of science, often demands a collective approach – a merging of disciplines to expand its scope of reason.

The collaboration of disciplines is likened to *bird verses cat* – often there's a disagreement and synchronism is scarce. Yet, as in the *real-world*, acceptance is key and affords synchronicity.

The disciplines that are comprised within the *framework* of *The Elite Research Method: a methodological framework applied in the real-world* are first, the rehabilitation or applied sciences, simplified by the health sciences with its varied terms, or health as its overarching discipline; and second, education and/or professional studies, with its own subdomain of pedagogical research, simplified herein as the education sciences.

Typically, the discipline of health is widely accepted as a complex science with its respective offshoots – utilising quantitative methodologies, those statistical in nature. However, the discipline of education is arguably far from this level of acceptance with its own respective derivatives –

utilising qualitative methodologies, theoretical in nature. One discipline is built on numbers and their application in answering questions. The other discipline is built on specific *real-world* answers that then draw on complex processes to draw out its corresponding data sets.

The immersion into either field is set with its own rules – one is centred on addressing questions with statistics, whilst the other is fixed on unravelling answers through its respective data sets. Both disciplines collect and collate their data respective to their discipline *rulebook* – that is, measures in place to adhere to strict research practices.

Fascinating *not* by chance, that one methodology is set on answering a question, whilst the other is drawn from the answer. Sound reasoning would demand the utilisation of both methods to have the utmost statistical and theoretical inferences for the sciences – yet seldom are used as a result of the need not being warranted.

Need stems from the *what* – that is, what is the research premise; what is the question that is seeking to be answered; or, what kind of data is needed to answer the research question and what method will permit the required data to be collected? It is key to note that a research question has multiple aspects – primary and secondary. Typically, the primary question will be addressed by its subsets (secondary) that ultimately answer the overarching question (primary). Equally, primary and secondary questions follow the identical process – the methodology employed to answer the *what*. However, there is a third methodology of choice.

A mixed method design encompasses both qualitative and quantitative research methodologies and draws from both processes. However, to warrant merit for its use, the primary question is required to have secondary questions that demand its use-case. For instance, a method cannot be chosen as a want – it must adhere to a need, therefore contributing towards the validity of the overall research.

It is well known in the sciences that one discipline will demand qualitative research methods to be employed, whilst the next discipline will demand quantitative research methods to be paramount. And whilst *birds and cats* in the *real-world* (recall the earlier analogy) scarcely form meaningful relationships – the sciences are no exception. Calling on more than one discipline is an arduous uphill battle in itself – meanwhile, employing more than one methodology is equally gruelling.

Yet, if your primary question has extraordinary implications for the *real-world*, an extraordinary fusion between disciplines and methodologies is a mere requirement – drawing on all the respective *rulebooks* afford to hopefully answer that *what* – there is never a guarantee. Science *is* science – a scientist cannot predict the statistical inferences or the theoretical implications without trying. However, they can pre-empt through a lived experience – a complete immersion, and hope for the best.

The Elite Research Method: a methodological framework applied in the real-world involved a lived experience. It set out to answer a primary question with extraordinary implications for the *real-world*. In order to succeed, a fusion between disciplines was a necessity. Recall, fascinating *not* by chance, that one methodology is set on answering a question, whilst the other is drawn from the answer. Thus, the reasoning is clear – whether a collective experience of arduousness and gruel-intensiveness, *The Elite Research Method* was born.

Encompassing both methodologies – qualitative and qualitative, whilst merging two sciences – health and education, *The Elite Research Method* was directly used to answer the doctorate titled *How is the V Position applied and communicated in the forehand groundstroke by elite players and coaches internationally?* Significantly, *The Elite Research Method* comprises of a direct extract from the one and the same.

The extraordinary of merit in answering the primary question was two-fold – the establishment of a *new* technical parameter (V Position), or technique, that contributes towards elite success, and *how* to develop the best players and coaches in the world. However, these answers are **not** captured inside *The Elite Research Method* – rather, its *framework* paramount to its unravelling. Noteworthy is the transitional use of the forehand groundstroke – a serial (tennis) skill comprising of a multitude of discrete skills. The formation of the primary question is in fact transitional – not only to the *real-world* in direct question, players/athletes and coaches, but its wider sub-groups. Synonymous to the opposing health *and* education disciplines, is the use-case for its transition in its wider application.

The Elite Research Method presents a *methodological framework applied in the real-world* – inclusive of its direct appendices that expose the arduous truth of gatekeepers in research, and the difficultly in cutting through fields ill-accepting of the other. Furthermore, a discrepancy exists in the use of the scientist's voice.

Qualitative research encourages a narrative to be explored between the scientist and the data – an invaluable recount of the process undertaken and the generated data. This process allows the data to come to life – a lived experience shared. To the contrary, quantitative research employs a strict and routine language – the written word is set on data-only with succinct detachment – opposing the applicability of experience in research. As a result, *The Elite Research Method* presents an incorporation of the two – a tiresome task when the criteria to be met is nothing short of prejudice.

Judgement has merit – it affords rigor. As arduous, gruelling and tiresome as *The Elite Research Method* demanded, an extraordinary framework prevailed. Before you is the direct process undertaken to achieve an extraordinary outcome. But is has **not** been left untouched. Every substantial process has been accompanied by a briefing – an overview of the subsequent context to keep you, the reader, on the map. For before you is your very own map, transitional in its use – for grasping its context, provides the essence for applying *The Elite Research Method*.

Map Briefing: An Overview

When undertaking a project, regardless of its shape, size or context, depth of thought is key in the decisiveness in the approach to be undertaken to ensure the project is accomplished. In the *real-world*, as in life, a process is necessary to be set out and followed to act as a map – a central reference point to keep you on track.

The initial section of *a methodological framework applied in the real-world* comprises of the research introduction that accounts for what has come before this area of work. As with research, symmetrical with stories, there is a juxtaposition between beginnings, middles and ends. The case here is a recollect of earlier accounts of what the research has so far achieved – being mindful that what lays before you is the juxtaposition of the middle. Therefore, there will be times throughout *The Elite Research Method* where the beginning and ends are referenced as a result of their significance overall – to the broader research focus; yet, when narrowing the focus to the *framework*, it is key to be mindful of the key difference.

The wider research compiled of an arduous undertaking of the sciences which was and is multifaceted in its entirety. However, it is this narrower focus – a direct extract of this work, accompanied by these briefings, that allow the story to further be unravelled. For to unravel is to break-down and share, ultimately allowing *The Elite Research Method* to be applied, again, in the shape, size or context of its next project.

It doesn't stop there, for this introduction is merely a stepping stone for the *framework* to come. The *why* behind the process is unveiled – the varied aspects of the *methodological* design and their use-case as per the *rulebook*, ensuring the duality of qualitative and quantitative methodologies exist in collaboration. Recall, merging the statistical and theoretical is one side, whilst accommodating one end (education) with the opposing (health) is another. To accommodate both sides, the nature of inquiry delves further to share this process.

Then comes the premise of the overall project – the central focus of the *framework*. No matter the overarching primary question, the secondary questions hold the utmost significance – these are what the primary question is reliant on to ultimately address its multidimensional variants. Further, these secondary questions directly tie back to the process itself. For without causation or mere warrant to build and ultimately apply a *framework*, it would be entirely different. It is these questions that inform the process of *The Elite Research Method* and ultimately its design that contributes to the *methodological framework* employed herein.

Research Introduction

In order to address the objectives of this research, the foundations need to be presented. These foundations comprise the methods that have allowed the data to be collected that produced the findings. It is important to reiterate at this point that the nature of this research bridges the fields of sports science and education to investigate the performance characteristics of the forehand groundstroke and how it is ultimately applied by elite players and communicated by coaches.

The multidisciplinary nature of this research, the diverse literature informing the theoretical framework, and the internationally dispersed participant pool, required a mixed methods approach, allowing quantitative data to be generated to report the frequency of the V Position, whilst also employing qualitative methods that enable coach perspectives on the V Position to be collated and analysed to define how coaches communicate the characteristics of the V Position (Denzin & Lincoln, 2011; Munroe-Chandler, 2005).

Cohen, Manion, and Morrison (2007), discussed that the integration of research methods is utilised when one method does not suffice in answering a research question. Termed "fitness for purpose" (p. 39), Cohen et al. and Trumbull and Taylor (2005a) have argued that a collaboration of qualitative and quantitative may be required to fit the premise of the research and effectively answer the research questions. Research methods can be categorised as qualitative or exploratory in nature; or quantitative and numerically infused (Trumbull & Taylor, 2005a). These differences will be further explored in the coming sections.

Considerations for designing research

The nature of this research integrates the fields of education and sports sciences. Thus it was appropriate to consider a wider range of tools that might be used to inform the research design. As a result, the research design follows a mixed methods paradigm, discussed by Trumbull and Taylor (2005a) as a model that allows the researcher to include the characteristics of qualitative and quantitative methodologies. Mixed methods was asserted by Cohen et al. (2007) as useful when one methodological paradigm and its typical methods do not entirely fit the research purpose. This mixed methods research design also accounts for my epistemological assumptions about the social nature of knowledge and ontological assumptions about the co-constructed nature of reality underpinning this research (Denzin & Lincoln, 2011). These encompass and inform the interpretation of how coaches and players come to know what they do and why they coach and communicate a certain way, drawing upon the strengths of qualitative and quantitative methodologies (Schoenfeld, 1989).

According to Trumbull and Taylor (2005b), a key variation in qualitative and quantitative research is that one tests a theory (quantitative), whilst the other develops a theory (qualitative).

Hennink, Hutter and Bailey (2011) argued that a traditional qualitative approach is utilised when the research requires theoretical (or narrative) and/or statistical data gathered through observation tools. Alvesson and Skoldberg (2009) suggested that research methods are complex in nature, and therefore the methodology can be customised to align with the purpose of the research. A mixed methods research design was defined by Cohen et al. (2007) as:

> "focusing on research questions that call for real-life contextual understandings, multi-level perspectives, and cultural influences, employing rigorous quantitative research assessing magnitude and frequency of constructs and rigorous qualitative research exploring the meaning and understanding of constructs, utilizing multiple methods, intentionally integrating or combining these methods to draw on the strengths of each; and framing the investigation within philosophical and theoretical positions," (p. 4).

This investigation required an approach that could combine methods to generate quantitative data around 'prevalence of constructs' (application of the V Position) as well as the meaning behind their application. As a result, a mixed methods design was most appropriate. The nature of quantitative or positivist research, and qualitative or interpretive research, is outlined in the following section.

The nature of inquiry informing the research design

A positivist research paradigm can be described as one that requires research to be conducted quantitatively in a social context (Hennink et al., 2011). Strictly positivist research also requires methodological approaches that limit or exclude the influence of personal experience on data interpretation. As a result, an interpretive paradigm is more suited to this research problem as it allows for methodological designs incorporating "interpretation and observation," and permitting personal experience to be an integral feature (Hennink et al., 2011, p. 15). Thus, an interpretive paradigm will enable explanatory research to unravel multiple possible explanations as to why and how the V Position is applied in the forehand groundstroke of elite tennis players.

A mixed method of inquiry was used which allowed the collection of specific numbers and statistics relevant to support the findings in a secondary context, helping answer the research questions (Trumbull & Taylor, 2005a; Webster & Mertova, 2007). A mixed methods approach was selected to inform the research design due to the epistemological and ontological assumptions that the researcher's insider perspective: the personal experiences that would act as a strength of the investigation, necessitating the inclusion of the subjective viewpoints.

Defined by Packer and Goicoechea (2000), "epistemology is the systematic consideration, in philosophy and elsewhere, of knowing: when knowledge is valid, what counts as truth, and so on," (p. 227). It raises many questions such as how can reality be known and what is the relationship between what is known and the knower (Vasilachis De Gialdino, 2009). In the sports coaching context, a coach's epistemological assumptions will determine what they consider to be acceptable evidence of performance, or ways of knowing (Vasilachis De Gialdino, 2009) or ontological assumptions about how one comes to know. Ontology, then, as defined by Packer and Goicoechea (2000) is "the consideration of being: what is, what exists, what it means for something—or somebody—to be," (p. 227). The epistemological and ontological assumptions shaping this research problem include that valid knowledge and therefore evidence of such knowledge is observable, experienced by performers, and calculable through systematic collection of data or results. An athlete's performance is observable, and how a coach (or researcher) interprets observed performance constitutes a valid body of knowledge, a KP, that can inform others. This valid knowledge, and valid way of knowing derived from interpreting athlete performance, is accepted by others who also share these epistemological and ontological assumptions about social reality; that notating and interpreting performance from those defensible notations constitutes valid knowledge: it is "what counts as truth" according to Packer and Goicoechea (2000, p. 227). Certain methodological approaches and related methods then follow the epistemological and ontological assumptions underpinning research. Due to the utility and pragmatism of a mixed methods design, four major data collection methods were employed: observation through notational analysis (Padgett, 2012), analysing texts and documents (including survey analysis), and interviews (Silverman, 2001).

According to Knudson and Morrison (2002), qualitative analysis is required to improve athlete performance. Qualitative analysis incorporates a variety of areas, specifically in concern to kinesiology, allowing the integration of biomechanics and coaching pedagogy to be interrogated through a qualitative lens, whilst the data itself can take both quantitative and qualitative forms (Edwards, 2009; Knudson & Morrison, 2002). It is on this basis that a mixed methods design was selected to investigate the research questions, encompassing both the subjective experiences of the participant group, and data of a statistical nature. The combined approach permits both interpretations of text and numerical data to co-explain phenomena, as opposed to using the more dominant, positivist approach undertaken in sport sciences, which insist on quantitative research methods that would exclude the participant story. Thus, methods of data triangulation were also employed, comprising of multiple methods that aim at contributing to the overall validity of the findings (Richey & Klein, 2007). The mixed methodology allowed for consideration of researcher and participant experiences, and statistical representation of the frequency and application of the V Position, something that would not have been able to be achieved purely with a qualitative or quantitative based method alone.

The research questions

To best answer the primary research question, six sub-questions were devised to shape the investigation. In order to answer Question One: *To what extent is knee bend prevalent at point of contact in the forehand groundstroke on hard court surfaces in elite tennis players?* and Question Two: *At what height is the ball at point of contact in the forehand groundstroke typically contacted on hard court surfaces in elite tennis players (in relation to hip height)?,* data had to be collected on the frequency of the identified movement characteristics of the V Position during competition. Observation data was created through notational analysis of unique performance criteria designed to account for the criteria and the variables of the V Position.

To answer Question Three: *What is the relationship between prevalence of knee bend at point of contact on hard court surfaces in elite players, and better performance (higher ranking), which would indicate if the "V Position" is prevalent in elite players?*, SPSS analysis of the observation data, combined with document collation, was required. Document collation took the form of player rankings as the benchmark of successful performance, focusing on players who were inside the ranking range of the top 200 on either the ATP or WTA tours.

To answer Questions Four, Five and Six, the methods used changed to fit the purpose of the question. The data were generated from the combination of survey and interview questions, whilst software-based and manual thematic analysis was used to suggest findings, drawing on both the conceptual and theoretical frameworks presented in previous chapters (Hennink et al., 2011). The survey and interview questions were designed to elicit data around coach communication and influences upon pedagogical approaches. To answer Question Four: *How do coaches' perceptions of their athletes' learning styles influence the approach taken by elite coaches in communicating lower body positioning (knee bend) in the forehand groundstroke, and how does the coach know if his/her approach is effective?*, data generated from the survey and interview method were interpreted drawing upon concepts derived from the literature to explain the different types of learning that impact on the development of a coach's pedagogy and how this pedagogy caters for the learning needs of the athlete.

To answer Question Five: *What are the teaching and learning approaches (or pedagogic approaches) used and communicated by elite coaches to teach their players knee bend at point of contact (the V Position) in the forehand groundstroke?,* the data generated from the survey and interview methods were analysed, focusing upon the pedagogic language used by elite coaches.

Lastly, to answer Question Six: *What other factors contribute or impact upon optimal coaching pedagogy in the forehand groundstroke at the elite level?,* a synthesis of statistical findings and the emerging narrative from interpretation of survey and interview findings was undertaken in order to identify factors that contribute to an optimal coaching pedagogy.

Map Briefing: An Overview

A central indicator of progress throughout the application of *The Elite Research Method* can be condensed into a simplified query: where are my secondary research questions leading, and are they answering the primary question of focus? Transitional in their use, secondary questions are key indicators of progress to the overall project. In this case, the project is the outcome of *Elite Research* and its subsequent findings.

In the *real-world*, an identical approach can be framed to consider the project of interest, then subsequently broken-down into the necessary secondary questions that will ultimately see the primary question – that of the project, be answered. Reflect on the intricacies of the secondary questions and their ultimate premise. Whether there is a dozen, or a half-dozen secondary questions, is dependent on the intricacies involved in the project, and/or the depth of focus to each respective question.

To take the next step in the *framework*, and to bring these questions to life, *The Elite Research Method* moves to the *how*. The processes in place to ultimately compile the data, or information necessary, identifies how these questions will be answered and permits the researcher, or project lead, to gauge its effectiveness.

It is important to understand that processes are open to change as a direct result of their responses. For instance, a pre-thought appropriate process may not gather nor provide the necessary data that was initially intended. In these scenarios, additional approaches are of merit to ensure the central indicator of focus, the primary question, is being worked towards, opposed to collecting data at the time that is insignificant.

To avoid null data sets, regardless of their form, *The Elite Research Method* collection methods are herein. All research and projects differ due to their respective variances and underlying connotations. Keeping this in mind, so is *The Elite Research Method* in its tangibility. Some projects may require equal depth, whilst others a pure qualitative approach – or to the contrary, qualitative design. Encompassed in all processes is the data collection method of choosing – whether mixed, or standalone.

The *methodological framework* incorporates varied aspects of the *rulebook* to provide unquestionable structure, and an open stringent process. The use-case of these methods and their undertone to the central research premise is herein shared – interviews, questioning methods and the structure, or semi-structure of the respective questions. Furthermore, the inadvertent use of external documentation and its role in contributing to the overall premise of the project, or research at hand.

It doesn't stop there. The *real-world* applicability begins with notational analysis and *how* it is an essential ingredient in *The Elite Research Method*. This form of analysis is coupled with analysis techniques compounded through typical reliability and video footage – applicable in this inference where *real-world* live depictions were crucial. Directly tied to and complementary to this *methodological framework*, is the use-case for the survey later shared.

The context to follow is shaped around the data the research demands, and the processes that in turn will afford the project its answers. A backwards model – from the answer sought, to the questions that will inform the answer – underpinned by the process required that will afford the data to be collected. After all, without the data, answers remain empty – questions simply remain.

Data collection methods

The following section details and justifies the selection of data collection methods employed in this study, in terms of how the combination of methods addressed particular challenges such as working with international and elite participants, and their suitability in answering the research problem. These data collection methods include qualitative interviews, document collection, observational analysis including analysis of video-footage, and the completion of the survey.

Interviews

Interview methods, as discussed by Moyser and Wagstaffe (1987), are typically included in research design involving elite participants, as the elites provide access to expert knowledge in their field. The expertise of elites was considered pivotal in answering the problem because of their experience working with, coaching and developing this level of expertise. According to Moyser and Wagstaffe (1987), elites are people who are at the top of the hierarchy in their own domain, although Hertz and Imber (1995) argued that elites usually have barriers that separate them from the rest of society (Hertz & Imber, 1995). Elite players and coaches typically have protectors, including personal staff, managers and tournament organisers, tasked with maintaining the distance between elites and the general public. This is why the survey method, discussed shortly, was also deployed online to efficiently collect the required information from geographically dispersed participants in a timely manner (Long, 2007). Purposive sampling method was also used to identify and collect participants on the basis of the expected/predicted knowledge and experience the sample represented (Long, 2007). Furthermore, participants were recruited through their association with each other. "Snowball sampling [is] a method that has been widely used in qualitative research [where it] yields a study sample through referrals made among people who share or know of others who possess some characteristics that are of research interest," (Biernacki & Waldorf, 1981, p. 141). This method of sampling was used to make contact with participants who had colleagues who would be equally as willing to participate in the research (King & Horrocks, 2010).

The interview method was originally conceptualised as a means to ascertain additional information that participants may have omitted from the survey. Interviews allowed for a wider context to be considered, incorporating a face-to-face approach or continued virtual (online) correspondence (Trumbull, 2005). Both of these forms were used as they permitted communication with participants to get a thorough understanding of their ideas about coaching the forehand groundstroke. The interview allowed the data collection to continue along a structured format, yet allowing other topics that the interviewee raised to be collated. Thus, as Trumbull (2005) explained, the interview process allowed for social interaction whilst collecting empirical data.

Questioning methods: the interview process

Qu and Dumay (2011) wrote that the research interview is a highly regarded and significant data collection method in qualitative research. However, complete objectivity by the researcher is not possible (Davies & Dodd, 2002). Davies and Dodd (2002) discussed that by not acknowledging the presence of subjectivity in the research process, it could have a detrimental effect. The role of subjectivity is furthered by Ortlipp (2008) when conducting interviews, making it clear that throughout the interview process, uncertainty is natural, and interaction between the researcher and the participant will always be questioned due to the notion of remaining as objective as possible, yet subjective as deemed appropriate (p. 702). According to Ortlipp, "researchers are urged to talk about themselves" (p. 695) in qualitative research. As such, reflexive practice throughout the research process "makes visible to the reader the constructed nature of the research outcomes" (p. 695) that are embedded in the decisions made by the researcher throughout the research project (Ortlipp, 2008).

The uncertainty and level of interaction through the research process was simplified by making visible and clear by employing consistent reflexive practice, through the use of a researcher journal, illustrating the visible role of the researcher in the overall research process. Davies and Dodd (2002) added that by reflecting on the research process, the researcher is given a voice to narrate responses that were expected or those that challenge the researcher's personal belief system (p. 286). These responses are in turn acknowledged and find a place within the data by allowing the researcher to be positioned subjectively in the equation. This is strengthened by the researcher's background and level of expertise in the topic. A key responsibility of the researcher, discussed by Qu and Dumay (2011), is to have an extensive knowledge base in their field or topic to allow the interview design, who to interview and what questions to ask, are organised and delivered to promote a rich data collection experience.

The interview itself requires considerable time and effort to ensure the questions posed are informed and will allow the researcher to collect the data necessary to their research objectives (Qu & Dumay, 2011). Qu and Dumay (2011) stated that "interviews provide a useful way for researchers to learn about the world of others," (p. 239) allowing the researcher to consider ideas and points of view that they would not normally think of, and in turn enrich the data context. Although there is no explicit way to conduct an interview, nor format that is going to be suitable for every qualitative interview (Qu & Dumay, 2011, p. 247), Turner (2010) refers to these interview types as informal conversational, the general interview, and the standardized open-ended interview

(p. 754). Qu and Dumay (2011) however referred to these differently, labelling the three central interview methods as: "structured, semi-structured and unstructured," (p. 239). A structured interview "is where the interviewer asks interviewees a series of pre-established questions" meanwhile only permitting "a limited number of response categories," whilst unstructured interviews take a more relaxed approach and "proceeds from the assumption that the interviewers do not know in advance all the necessary questions," (Qu & Dumay, 2011, p. 244-245). However, a mix between structured and unstructured is semi-structured – the most dominant approach in qualitative interviews (Qu & Dumay, 2011). Semi-structured interviews are prepared, although allow the freedom to probe the interviewee to deepen their responses (Du & Dumay, 2011, p. 246).

The interview findings should be viewed as an abundance of data collected to address the interviewer's research objective. Post interview, the researcher is left with an endless supply of qualitative data that acts as a central agent in answering the research questions (Turner, 2010). Turner writes that although the data has been collected, a daunting process awaits, requiring "the researcher to sift through the narrative responses in order to fully and accurately reflect on... all interview responses through the coding process," (p. 756). Additionally, collating data poses another challenge, as Turner notes that interviews are predominantly "coupled with other forms" (p. 754) of qualitative research methods to provide an overall informed data set. In order to understand the data, Turner noted that themes and codes are compiled to separate the data into "sections or groups of information" (p. 759) that represent consistent or reoccurring phrases or ideas shared amongst the participant pool.

Both open and closed ended questions were included in the interview process; closed questions allowed for responses linked directly to priorities identified in the literature review (that is, in answering gaps in existing literature) whilst open ended questions allowed coach participant expertise to shape the nature of responses, allowing for emotion, perception and context (Trumbull, 2005). The interview questions further asked the participant group to confirm their views on the movement characteristics of the V Position, whilst a better understanding was formed of what informed their coaching pedagogy and how the participants communicated these elements.

Initial, structured coach interview questions

The semi-structured interview questions were founded on the participant's response to the survey (Appendix A) that he/she completed. Comprising five open-ended questions, the questions were aligned with the order of the questions in the online survey, allowing the respondents to provide further comment about them. These questions are now displayed in full.

1. In regards to the survey completed, can you please further your answers in relationship to the first question group and the previous players coached and the duration you coached the respective players?

2. In regards to the survey completed, can you please further your answers in relationship to the second question group and your coaching pedagogy and where it is derived, along with how you communicate lower body movement with your player and what you have found to be most effective? In addition, can you please explain your answer in regards to knee bend in the groundstroke and if it has a place?

3. In regards to the survey completed, can you please further your answers in relationship to the third question group and the location of where the groundstroke should be contacted and the generation

of power? In addition, can you please explain the use (or not) of knee bend in the groundstroke and the required stance of the player when making contact?

4. In regards to the survey completed, can you please further your answers in relationship to the fourth question group and the knee bend and contacting at hip height, the relevance, if any, and how you approach this? In addition, can you extend on your response to coaching discrepancies (if any) and how you approach coaching your player in respect to their lower body movement?

5. In regards to the survey completed, can you please further your answers in relationship to the fifth and final question group and the understanding of your player and how you know your coaching pedagogy is appropriate?

Adapted, semi-structured coach interview questions

The interview questions were consequently adapted to account for interview participants who had not completed the survey. This required a minor adjustment of the initial interview questions to ensure any reference to the survey (as worded in the original questions), which may have caused confusion, was avoided. These questions are now displayed in full.

1. Can you please give a summary of your coaching experience: recent players' rankings, your achievements with them, number of years coaching each player, any accreditation you have etc. Whatever you would like to tell me about yourself! All coaching profiles will remain anonymous in reporting.

2. How would you describe your coaching pedagogy (for example, your coaching approach to improve player learning) and how/where it has developed? What

should a desirable lower body position and movement look like in the forehand groundstroke (for example, if knee bend is necessary, and if so, how deep is the knee bend as per glossary description)? Can you elaborate on how you communicate/teach the player to obtain the desired lower body movement in a forehand groundstroke (for example, particular concepts, cues, questions or activities that require them to think about and achieve the right movement)?

3. How important do you think knee bend is at point of contact in the forehand groundstroke? Why or why not (if important, please specify slight or deep knee bend)? How does the height of the ball upon contact, degree of knee bend and power generation in the forehand groundstroke work together in your opinion?

4. At what height is it most effective to make contact with the ball in the forehand groundstroke in relationship to the body (for example, above hip height, around hip height, or below hip height)?

5. What kind of feedback do you look for to see if your coaching approach to improve player learning (pedagogy) is working well and suits the player?

Document collection

Document collection can include a myriad of texts (Jupp, 2006). The documents required and therefore collected in this study came in the form of the player rankings which are available to the public and sourced from the WTA and ATP tours. Additionally, video-based images were generated using Kinovea software: a biomechanical based software enterprise, freely available to the public.

Notational analysis and typical reliability

Dogramaci, Watsford and Murphy (2011) noted that subjective notational analysis, a systematic approach to collecting observational data (Hughes, 2002), is an effective data collection tool for analysing and recording athlete performance, movement patterns and technique changes. Discussed earlier as a "valid form of knowledge" (p. 852), Dogramaci et al. (2011) highlighted not only the benefits of subjective notational analysis but a preference for it over other performance tracking measures. Although laborious, Dogramaci et al. suggested that not only is it an effective data collection tool, but produces high quality data irrespective of the subjective role of the researcher. Callaway and Broomfield (2012) also claimed that notational analysis is a useful tool when analysing athlete technique or movement patterns. Dogramaci et al. furthermore argued that it is an accurate measure of data in court and net games, like tennis, due to the "short distances and changes in direction" where the human eye is capable of recording and analysing these movement interactions (p. 859).

Notational analysis has been applied to the sport of tennis previously. For instance, Gillet, Leroy, Thouvarecq, and Stein's (2009) applied this technique on the tennis serve for 116 men's singles matches. In comparison, the notation used in this research centred on 123 elite tennis players, both male and female, with 11 values considered (with 121,770 individual notations). The observation technique was used, and has been used in other sports with similar movement sequences. For instance, Lupo et al. (2011) used notational analysis to look at technical and tactical performance indicators in water polo, whilst James, Mellalieu, and Jones (2005) applied the technique to identify position-specific characteristics in rugby union, demonstrating that notational analysis has become a common tool for coaches and/or researchers to analyse player/athlete performance in order to deliver game or match specific feedback. Clemente, Couceiro,

Martins and Mendes (2012) also applied notational analysis techniques to provide best practice indicators in soccer, allowing coaches to construct game-specific feedback, whilst Angel, Evangelos and Alberto (2006) used the observational tool in basketball to identify defensive systems, and Ciuffarella et al. (2013) employed notational analysis to record a variety of variables in the serving techniques of volleyball players. To illustrate the input of quantitative research methods when generating results, Kwitniewska, Dornowski, and Hökelmann (2009) described the use of SPSS, among other software tools, in the analysis of rhythmic gymnastics. Additionally, Araya and Larkin (2013) also used SPSS to generate statistics on the notational analysis of central performance indicators in the English Premier League.

Reliability measures for subjective notational analysis can take the form of intra-rater reliability or inter-rater reliability. Intra-rater reliability is where data is collated and observed by one researcher, with the reliability arising from the singular, consistent analysis by one experienced person, knowledgeable in the performance being observed with minimal interference (Viera & Garrett, 2005). McGarry and Franks (1996) noted than at the elite level of sport, elite athletes are conditioned to perform their strokes/shots a certain way, and therefore when considering the characteristics of these shots on multiple occasions throughout a game/match, the way the shot is struck and/or where the athlete moves to make contact with the shot, becomes predictable (p. 451).

Cushion, Harvey, Muir and Nelson (2012) noted that when establishing reliability through notational analysis, these forms of analysis are restricted by design in what they can measure, as they usually disregard context. Inter-rater reliability can be employed, which requires more than one person to be involved in the notational analysis process, independent of the researcher's. These separate results can then be compared to the primary researcher's notations and a measure of reliability can subsequently be reported (Mooi & Sarstedt, 2011). Callaway and Broomfield (2012) added that although in "reliability studies the actual results are not known...[these are then] compared to" additional measurements (p. 295).

McGarry and Franks (1996) noted that when considering invariant behaviour at the elite level, performance characteristics of a previous shot/outcome become more predictable. At the elite level of sport, these athletes are conditioned to perform their strokes/shots a certain way, and therefore when considering the characteristics of these shots on multiple occasions throughout a game/match, the way the shot is struck and/or where the athlete moves to make contact with the shot, becomes predictable (McGarry &Franks, 1996, p. 451).

The observation analysis techniques and specific notational tool will now be discussed and how subjective notational analysis was employed throughout the data collection process.

Observational analysis techniques

According to Ronglan and Havang (2011), the process of observation is a necessary action in order "to grasp and understand the social reality we are a part of... it means that we have to separate something from something else, otherwise everything merges into an incomprehensible chaos," (p. 81). To answer the research questions, as discussed by Hennink et al. (2011), observation of player performance in 'real life' match-play context allowed for recording from a non-intrusive distance (in-person and via recorded footage) and later data analysis. Observation was selected to overcome the foreseen challenge of getting in close contact with elite players (Denzin & Lincoln, 2011).

Notational analysis, a form of observation specific to sporting disciplines, allowed for the analysis and subsequent observation of movement patterns and player/athlete performance (Hughes, 2002; Hughes & Franks, 2004). As discussed by Taylor, James, and Mellalieu (2005), notational analysis is a tool that allows coaches, or researchers, to gather objective data and then use it as a feedback mechanism on player/athlete performance. Notational analysis techniques were used to document the frequency of observed variables of the V Position, having the capacity to consider the forehand groundstroke, player positioning (balance) and the level of knee bend applied at point of contact, as well at the height of the ball when the player executed the stroke (Hughes, Hughes, & Behan, 2007).

Notational analysis provided a statistics-based end result which is in turn explained in combination with other qualitative findings (Hennink et al., 2011; Knudson & Morrison, 2002; Trumbull & Taylor, 2005a). The use of SPSS or similar software is a common tool used to generate statistics from notational analysis in order to provide cumulative results (Mooi & Sarstedt, 2011).

Through the notational analysis used in this study, two forms of observation occurred: in-person (live) observation and video-based observation methods. Both forms of observation required an observational chart to be designed which is an adapted model derived from Elliott (1991, 2002), Schonborn (2000) and, Elliott, Fleisig, Nicholls and Escamilia's (2003) previous blueprints. This chart accounted for characteristics that influence player positioning at point of contact in the forehand groundstroke. Video-based observations were used for two reasons: one, to ensure ample data was available, beyond what one researcher could reasonably collect from live performances, and, two, so that there was a record of players who may not have been available to observe in person due to multiple matches occurring at once in the same tournament.

To provide a visual representation of the overall results, the use of Kinovea software was incorporated, providing further illustration of the movement characteristics of the forehand groundstroke. Software with similar features to Kinovea is widely used in research for video analysis of movement. For instance, video-based software was used by Kuntze, Mansfield and Sellers (2010) for notational analysis techniques in the analysis of lunge frequency in badminton, whilst Kwan and Rasmussen (2010) applied a data acquisition system in the analysis of elasticity in badminton rackets during a stroke. The use of notational analysis allowed the results to be recorded against the personalised criteria, whilst Kinovea software provided a visual representation of 'what' was being recorded. To generate statistics of the observation process, SPSS allowed each variable to tell its own story whilst also considering the variables of the criteria together to indicate the prevalence of the V Position.

Video footage

The analysis of video footage in this study served two purposes: as a form of reliability to enable inter-rater reliability measures and to broaden the sampling of the study to analyse matches that, with tournament timetabling constraints, were unable to be attended in person. This method of data collection was complementary to real-time notational analysis.

The data collection process was timed in line with the Australian Summer of tennis: a time of year when tennis is free-to-air in Australia and broadcast for the duration of the tournaments. This time falls between late December through to late January each year, and is a result of the format of the WTA and ATP tours' yearly schedule. Most significant is that the Asia-Pacific Grand Slam is held in Melbourne, Australia every January; one of the biggest four tournaments annually (there are three other Grand Slam tournaments, held in Europe, England and North America respectively).

To ensure ample footage was collected, the Australian Open (the Grand Slam) was recorded as well as the lead up events, specifically the Hopman Cup (held in Perth, Western Australia) and the Brisbane International (held in Brisbane, Queensland). Both the Melbourne and Brisbane tournaments were attended to record in-person observations.

Survey development and explanation

The utilisation of the survey as a data collection method permitted questions to be asked to a wide, internationally based audience whilst controlling the nature of the question, intended context and response. The survey included closed questions in a multiple-choice format, and open-ended questions, with text box answer options, allowing space for individual opinions and explanations. The survey tool also ensured participant information could be kept anonymous if preferred, in accordance with ethical guidelines, or could elicit identified information if the participants chose to disclose it. Privacy was an important element to ensure, due to the potential participants' status.

The suitability of the survey and its questions were expert peer reviewed and trialed amongst a select few tennis coaching colleagues, and approved by the university for distribution. As a result of the timing of University ethical clearance and the imminent Summer of tennis schedule in Australia, the opportunity to trial the survey questions for a longer period, and with more participants was not logistically possible. Thus, expert feedback in the form of supervisor guidance was drawn upon to refine the questions. Such feedback included simplifying the wording of the questions and the incorporation of more open-ended response options.

In order to answer the main research questions, the survey questions collected data specific to coaching pedagogy, athlete management, the coach-athlete relationship, feedback mechanisms and language used, movement analysis and specific coaching of the lower body and contact point during the forehand groundstroke (see Appendix A). The personal and professional coaching experiences of the participants were specifically sought, to characterise the sample population, which would become important later in generating descriptive statistics. All participants were requested to complete the questions online, which was securely stored in the University Research Survey Centre, and enabled efficient and accurate transfer of data into SPSS. Below is an example of a survey question, with an explanation of its purpose in the survey (Question B3):

> B3 What forms your coaching pedagogy? (*Open ended response*)

This question afforded the opportunity for the participant to express where his/her coaching pedagogy originated and why he/she has chosen this path and why he/she coached in the format which he/she believed best. The question was very important to this investigation as understanding the basis for these participants' coaching pedagogy was quite significant given the elite level of players the coach(s) were currently or had previously worked with. An example of a multiple-choice question, with optional open-ended response, is provided below (Question B4):

> B4 How do you communicate lower body movement to your players for the groundstroke? (*Multiple choice*)
> - a) Get Low
> - b) Bend
> - c) Bend your knees
> - d) Lean forward
> - e) Lower yourself to the ground
> - f) Drop
> - g) Leg Drive
> - h) Drive through your legs

      i) None of the above (please specify what you would say)
      j) All of the above
      k) Other (please specify what you would say)

Additionally, to identify if the coach believes he/she uses effective communication strategies relevant to, or met the needs of, his/her player, Question E1:

      E1 Does your player understand what you say and take it in the correct context? (*Yes/No*)
      E2 How can you tell if he/she understands? (*Open ended response*)

The answers to these example questions could illustrate how these participants developed their coach-athlete relationship and subsequent pedagogy. According to Purdy and Jones (2011), an effective coaching pedagogy incorporates effective coach-athlete communication and positively impacts upon athlete performance. A coach can be identified as having a good relationship with his/her athlete when he/she knows when the athlete does and does not understand something and, importantly, knows how to communicate with the athlete when they do not understand.

Map Briefing: An Overview

Qualitative and quantitative methods of inquiry extracted from the *rulebook*, were customised to meet the demands of the overarching research premise. The applicability of these methods of inquiry, although brief in some instances, see document collection, extended to the more in depth, from notational analysis to typical reliability.

Length of extrapolation aside, typical project processes employed are customary and thus do not necessarily demand explanation. To the contrary, the non-typical and those rarely seen in *frameworks*, or in the company of further non-typical processes – the relationship between the typical and non-typical and their rare use-case in *frameworks*, yet alone in collaboration, demand clarification. After all, if it is yet to be seen, a firm premise is essential for its underlying reasoning to be impartial to the *rulebook*.

The next step in *The Elite Research Method* leans towards quantitative methods of inquiry – the statistical proponent of the *methodological* design. The design itself is *not* shared – for every research demands a model to work off, from and towards, *without* bias. Therefore, this particular design has been left empty – in its place is something much greater, if not the central index of *The Elite Research Method*.

Intentionally muddled in between the dialogue is a map much greater than any design that has come before us in *The Elite Research Method* – tied directly to the underpinned theoretical *framework*. For every project has an underlying measure – a causation behind its execution. In research, this is surmounted by the theoretical *framework* put forward – either directly asserted and *proven* by *a new theory* – an addition to the academic literature as the world knows it to be – the climax of research. Or, a contributing theoretical *framework* that may be disproven through the research, and/or its implications are true, yet insurmountable in the grand scheme of the research premise. In other words, a *framework* is either newly designed and ultimately established as a new theory through its climax – or, a theoretical *framework* utilises current theories and extends the literature in another direction.

Fortunately, in the case of *The Elite Research Method*, multiple theories were incorporated in the theoretical *framework*. Utilising the literature of the old (former), the theoretical *framework* that underpins the *methodological framework* herein, was proven by utilising *The Elite Research Method*. Both theories were respective in their own right – one being quantitatively based (refer to the V Position throughout) whilst the other is qualitatively based – as shared in 'The Science of Elite Performance: The World Awaits – *an infusion of science, education and communication: a complete learning structure on how elite coaches and athletes become the best in the world*': the Optimal Performance Theory. The Optimal Performance Theory is denoted as "the learning process and key communication exchange, informing The V by Dr. B ©" or herein referred to as the V Position (Berge, 2017, p. xi).

It becomes evident the more either theories are discussed, the complexities behind these are quite significant. The qualitative theory that was the climax of this *framework* in the Optimal Performance Theory, whilst the quantitative *real-world* application in the V Position, or "The V by Dr. B ©" (a technical parameter) was the overarching muse of the *methodological* design to ultimately be proven as truth (Berge, 2017, p. xi). However, how these outcomes tie directly back to *The Elite Research Method* and its map remains to be explained – until now.

The map of *The Elite Research Method* is used to pull together the concept (theoretical *framework*) and the application (project premise) to ensure all angles are covered – pulled from the *rulebook* with the respective *new model*. This map is at the centre of *The Elite Research Method* presenting the *framework* itself employed to gather the *methodological* datasets, respectively.

Reasoning becomes clear through the map and its modelling of how both the qualitative *rulebook*, and quantitative *rulebook*, merge in the *real-world*. To deliver an emergence of two divergent fields, health *and* education, the map provides a semblance to data and how it can be used in the *real-world*.

To pre-empt the map's premise, *mapping the data collection tools against the theoretical framework*, the forehand groundstroke is a technical parameter that was not only used as a key performance indicator of the V Position, but furthermore, as a transitional technical parameter to step outside of the sport (tennis). For point of reference, the forehand groundstroke is a serial skill in tennis, whilst the V Position comprises of varied discrete skills. Therefore, the criteria, observational and the selection for analysis, coupled with the exclusion criteria, are key stringent processes put in place to ensure the quantitative measures of the map were not only of equal substance, but placed against stricter precedents to move beyond the quantitative *rulebook* to enforce measures beyond doubt.

With any project, there is a central proponent of focus – or tool that is undisputed. Although there may be vast, most concrete research premises are narrowed down significantly to a core focus. The same applies for projects in the *real-world* – that central parameter guides the projects direction and its underlying processes. As evident by now, *The Elite Research Method* used the forehand groundstroke.

As identified in the map to come, although the forehand groundstroke is the *glue* that holds the *framework* together, it is by no means the end-point. The uncovered *new model*, *new theory*, and *new parameter* shared, are consequential end-points – whereby the forehand groundstroke is a 'vehicle used to travel the map'. The *methodological framework* affords these multi-climaxes as transitional in their application – a result of ensuring the parameter of focus was equally transitional in its use and wide-spread in the projects varied datasets.

Observational tool and criteria

The focus of the observation was to record the lower body movement characteristics of the players during the forehand groundstroke. The tool developed was informed by the four tasks identified by Elliott et al. (2003): preparation, observation and evaluation in respect to point of contact and the role of knee bend. The criteria were crafted to focus attention on the movement characteristics of the forehand groundstroke, and the works of Schonborn (2000), Elliott et al. (2003) and Elliott (1991, 2002) were used to design the observation chart. Schonborn's work was selected because of the clear depiction of relevant movement patterns, that is, Schonborn had reported on the presence of knee bend in the forehand groundstroke and also noted the importance of height of point of contact. The observation chart was also influenced by Elliott et al.'s design, as it offered a breakdown of the required components of the forehand groundstroke, allowing multiple characteristics of the forehand groundstroke to be considered and analysed through the one tool. For example, the characteristics of the forehand groundstroke of interest, which formed the notational criteria, are balance, knee bend and height of point of contact. These three combined criteria have not been considered in one design prior to this study. The previous work of Elliott (1991, 2002) and the break-down of the groundstroke and tennis serve mechanics acted as catalysts for this design. The following section considers the notational analysis and the selection criteria for the observation method.

Groundstroke selection criteria for observational analysis

The observational process of identifying the performance variables/characteristics required a specific categorical data set. A categorical data set allowed the inclusions and exclusions of the forehand groundstroke to be identified, which included all forehand groundstrokes played within the set boundary (illustrated in Figure 1 shortly).

The notations considered three criteria: balance, presence of knee bend and height of point of contact, with a further inclusion criterion of shot outcome. For the shot to be notated, it had to be classified as a forehand groundstroke, meaning that it fell within certain boundaries of the court and was not played in a compromised position; for instance, falling over or hitting off one leg. The baseline boundary for a forehand groundstroke to be notated was set 1.5 metres in front of the baseline (as illustrated in Figure 1) as contacting the ball beyond this point was considered to be an approach shot, altering the player's action and therefore, no longer a groundstroke. Shots played 2.5 metres or more behind the baseline were also excluded due to their defensive nature. These margins were set to focus on a typical forehand groundstroke that is characterised by baseline-play: balls contacted further back not only compromise a player's balance, but also categorise the shot as defensive, as opposed to offensive (attacking). Closer to the middle of the court is where a groundstroke transitions into an approach shot.

The boundaries in place limited the observation and subsequent notations to only those forehand groundstrokes that conformed to the set guidelines, excluding for instance, players falling over as they contacted the ball, regardless of whether it was otherwise inside the designated boundary area. All shots that were counted followed these strict guidelines to ensure the reliability and validity of the notational analysis, and that the player's technique observed was a consistent representation of the movement characteristics of their forehand groundstroke (see also Figure 2 for boundaries).

Figure 1   **Baseline Boundaries for capturing the forehand groundstroke in action—depicted as a metre guideline.**

Figure 2   **Baseline Boundaries for capturing the forehand groundstroke in action—depicted as a full court with the player's starting positions (serve/return).**

The speed of the ball was not included in the notations. Such ball speed generated by the opponent may be considered to be a central indicator to the calibre of shot and of course may limit the options for return. However, it is claimed here that it is not the opposition's ball speed or quality of shot that controls a player's ability to execute the forehand groundstroke, but at the elite level of play, the player's ability to move towards, or away from the ball is key to successful shot execution. That is, a player's ability to return the oncoming ball is not dictated entirely by the pace of the ball. The end result, the return, is reliant on player anticipation, fitness and speed around the court, and essentially their own positioning, where they have *decided* to make contact with the ball, which in turn is responsible for the height of point of contact of the resultant forehand groundstroke. Therefore, the application of the V Position is proposed to be less affected by ball speed, than the purposeful lower body positioning of the player, altered according to the decided height of point of contact. This is due to the calibre of the athletes as elite athletes have the capacity to readily move towards the ball whilst maintaining balance and preparing for the oncoming ball. Forehand groundstrokes played where the athlete compromised their balance and technique in order to keep the ball in play were also excluded. Balls that were contacted below the athlete's knee region, about to bounce for the second time, were also excluded due to their defensive technique, with compromised backswing. Sometimes, all criteria were met except for clear balance; these shots were included in the notation but described as unbalanced.

It was determined that the number of forehand groundstrokes recorded and notated should be 30 over an entire match, that is, that a match was observed until 30 forehand groundstrokes were recorded. If they only reached 28 and the match was completed, the match (and player) was excluded. A match was defined by as "the best of 3 sets (a player needs to win two sets to win the match) or to the best of five sets (a player/team needs to win three sets to win the match," [ITF, 2011, p. 6]). In Grand Slam events, males are required to play the best of five sets, whilst in all other events, and for female matches, the best of three sets applies. The 30-ball criteria was developed as it may be argued that if only 10 forehand groundstrokes are played in an entire match, it would not be a fair or true depiction of how a player consistently performs the technique. Thirty forehand groundstrokes would enable a better representation of the movement characteristics and was set for several reasons. One, as a result of much playing and coaching experience, the researcher has noted that on average, 30 forehand groundstrokes is a good indicator of a dominant shot selection throughout a match. Two, it was essential to ensure the groundstrokes notated were within the one match, meaning that conditions and context of their performance was not going to change. Therefore, setting a much higher tally of 100 balls, for example, would be unlikely to be achieved in one match. Finally, ensuring the observations were achievable within *part* of one match pragmatically allowed the researcher to observe as many live matches as possible, after notating 30 balls, rather than being locked into one match in its entirety at the expense of watching more players.

To decide what was 'enough' footage, the qualitative literature on notational analysis and the ideal number of observations is unclear and scarce. Hughes and Bartlett (2008) however recalled that "parameters and movement variables can be considered as performance indicators" (p. 740) and noted typical tennis performance indicators as "winners to errors ratios, shots/rally, quality – service/return" among parameters and/or variables that can be considered when conducting notational analysis," (p. 741). In addition, Hughes and Bartlett discussed the quantity of shots as a measurable variable, and that the use of ratios, contrasting variables of the criteria against others, are an effective performance measure providing that the provided ratio is out of a specified total of recorded variables. Shot execution is one such factor discussed by Hughes and Bartlett in net and wall games that can be calculated in ratios.

After tallying 30 forehand groundstrokes, the notational analysis considered the movement characteristics of the forehand groundstroke that represent the V Position's features: the presence of knee bend and the height of point of contact, which formed the mixed criteria for the analysis. The occurrence of one movement characteristic or criterion was contrasted to the others in the form of ratios, but each were noted individually to start with, ensuring a clear and concise depiction of each. These processes were also consistently applied in video observations. Video observations were planned to add to the number of live observations, rather than double up on live observations, serving to provide a close to 50-50% proportion of live to video-based notations, and a broader participant pool. For example, most matches were filmed by commercial television stations on centre court, providing recorded footage of players with higher rankings, whereas live matches were often lower ranked players on outside courts. A factor considered when generating the data set was that some matches were viewed live *and* also via recorded footage.

The exclusion of opposition quality in the selection criteria

The opposition quality was discounted by the measures enforced by the notational analysis and the boundaries set. These include the baseline boundaries, ensuring a player's movement characteristics were only recorded if they made contact with the ball in the strict forehand groundstroke hitting zone, not outside these boundaries. The player was required to be in a position to hit the ball, that is, they are not falling over or out of position (ensured again by the boundaries in place), and that they were hitting a forehand groundstroke, no other shot that has similar characteristics to the groundstroke, including discounting the use of a slice shot or 'tweener' (a forehand groundstroke played through the legs) that may be also struck in the set boundaries on the forehand side. Therefore, the boundaries set have been stringent to ensure that only the forehand groundstroke was notated and that, irrespective of opposition quality, the forehand groundstroke was primarily defined by where the ball landed in the court and from where the ball was contacted.

Statistics from both the survey results and notational analysis of observed performance were generated and analysed using SPSS. The two categories of statistics, descriptive and inferential, were applied to either summarise data or extend beyond the data itself and explain its implications and/or reasoning (Trochim, 2006; Lund & Lund, 2013). Pagano (2006) described descriptive statistics as being "concerned with techniques that are used to describe or characterize the obtained data," (p. 10). Inferential statistics according to Pagano are differentiated from descriptive statistics as they "involve techniques that use the obtained sample data to infer to populations," (p. 10). Both categories of statistics aligned with this study's research premise where descriptive statistics were necessary to show patterns that might emerge from the data (Lund & Lund, 2013), and inferential statistics were used to make judgements of the probability that an observed difference between participants was a reliable one or one that might have happened by chance (Trochim, 2006).

Mapping the data collection tools against the theoretical framework

The data collection methods outlined above followed a mixed methods design, informed by the theoretical framework. Figure 3 illustrates each individual data collection method and how it was used to answer the research questions. Each concept bubble in Figure 3 serves to indicate what was found by what data collection method: survey (green), interview (orange), notational analysis (purple: live, and blue: video), and document collection (pink) that included images generated from Kinovea software. For example, the survey (green) was used to collect data on how the coach learned, opinions on height of point of contact and knee bend in the forehand groundstroke, how coaches communicate movement, how these coaches deliver feedback, and what they believed contributed to, and was a part of, their coaching pedagogy. The video observation (blue) provides an example of what was collected through the notational analysis, the three criteria of the V Position and the inclusion criteria, shot outcome.

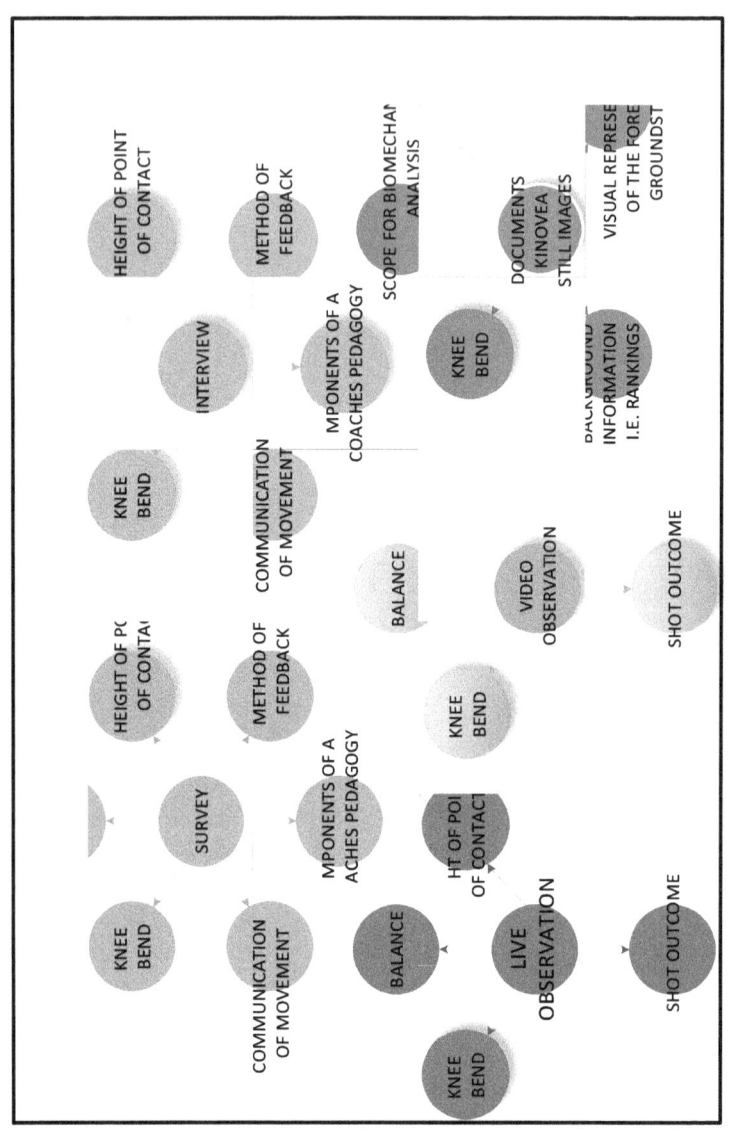

Figure 3 The incorporation of the framework through the data collection process: the five methods of data collection and the type of data that was generated through these methods.

In Figure 3 above it is clear that the survey and interview methods generated data across more themes than observation and document collation. This was made possible by the extensive range of questions in the survey and the allowance for open-ended responses, aimed at collecting data on participant understanding of, and communication of the V Position's criteria and its subsequent variables. Further, the interview questions allowed for open-ended responses which in turn permitted participants to provide as much information as they were willing to provide or share. The data arising from the observation was intended to focus on the forehand groundstroke and its use in the world's best players, and if the criteria of the observation tool were identified. Document collation was then intended to focus on participant profiling, such as ranking, and the use of images to support the findings of the notational analysis. The research action steps will now be outlined to illustrate how the investigation unfolded.

Map Briefing: An Overview

Identifiable through the data collection process, no less than *thirty* distinct forms of data were named in the *framework* that incorporates the five respective methodologies. An approach that set the groundwork for merging qualitative and quantitative forms of analysis to work in harmony *in the real-world*.

To initiate the *framework* and commence its *real-world* application, the *methodological* tone moves towards the action steps required to bring *The Elite Research Method* to fruition. Encompassed and presented is a journey that culminated over *two years*. The most arduous processes over this two-year period centred on notational and statistical analysis coupled with the collection process itself. It is noteworthy to share that throughout the research journey no less than four months consequentially saw these action steps come to a halt wherein the *framework* sat idle. This is directly referred to in the *foreword*.

Through the research process, detours are encountered. As in life, when planning the journey ahead, expect the unexpected. This expectation of a 'curve ball' affords the research journey to accommodate the *real-world* and thus be a mere extension of reality.

Directly tied to the processes herein, 'curve balls' were frequent with rarely a pre-planned process going ahead without a push back. Through the development of *The Elite Research Method*, and the already uphill battle set in place – this was merely the beginning. Although described earlier as arduous and gruelling – recall the extraordinary. Herein rings true. In order to achieve what is yet to be achieved or yet merely accomplished, extraordinary measures were of demand. A character trait demanded by and through *The Elite Research Method* is persistence, and akin to stubbornness, the will to see the project through. Daringly said, *The Elite Research Method* would not be nor have achieved what it has, without this stone-set character edge. Heave warning, when undertaking such projects be sure to equip your research journey with this form of undertaking, and sheer commitment.

This brings us to what *The Elite Research Method* unfolded – beginning with the ethical process. All research, projects too, have a governance system that approves the overarching premise to be undertaken. Prior to approval, merely the subsidiary processes can be undertaken. However, conditional of ethical approval, the primary components, largely the data collection process, cannot be initiated – until after approval.

Dependent on the severity of the project – the involvement of others, or not – with varying degrees of governance merely in this domain, ethical clearance is a feat in itself. Appendix B shares the formalities of ethical clearance in a University environment when conducting research at the highest stage (doctoral) and the amendments required to conform to the respective governance rules when undertaking *The Elite Research Method*.

Not the most titillating aspect of the research process, ethical clearance affords a rudimentary alignment for the conduction of research itself, and without it, the *rulebook* may be put into jeopardy. Thus, although mundane in the overall scheme, it is by far a necessary step throughout the research journey.

Directly tied to the ethical process is the data collection process itself and the involvement of others. *The Elite Research Method* demands anonymity for all involved due to the elite nature of the participants. This included, but was not limited to, current *and* former Grand Slam Champion players, and coaches. In essence, this depicts both players and coaches at the highest echelon of the sporting spectrum in their respective sport. Players who have won a Grand Slam Championship, and coaches who have coached players who have gone on to win a Grand Slam Championship. As a result, the rigid structure of the data collection process was well versed, and a key criterion to the overarching recruitment process.

Never before has the gatekeeper process of *The Elite Research Method* been shared. That is, up until now. A key proponent of the data collection process, the negotiation of gatekeepers is critical. In this inference, gatekeepers are persons who stand in the way of recruiting participants, or formally block all roads of communication, *and* access, between the researcher and/or project lead to those your research premise is dependent on. Essentially, gatekeeper's cut-off access to the participants your project is dependent on – and although ethical approval has been granted, these gatekeepers go out of their way to make this process the most difficult task of them all. Arguably, in the case of *The Elite Research Method*, it went against what was formally asked – thus the political landscape that followed required sheer gallantry to overcome and prevail.

Compounded by its length, Appendix D is a tale of the gatekeeper process via its very email exchange. To incorporate all emails would be a gargantuan task – and equally a text in itself. Therefore, the collection of email exchanges shared inside Appendix D is in fact a replica from the surmounting doctorate *The Elite Research Method* has largely been extracted from. Fair warning – a cross between mundane and the tireless process that followed. However, the final 'hurrah' was in the climax afforded by *The Elite Research Method* and its action steps that surmounted in the extraordinary. Recall, 'in order to achieve what is yet to be achieved or yet merely accomplished, extraordinary measures are of demand'.

Research action steps

In order to collect my data, I followed the road map informed by the theoretical framework developed (exclusive). There was a systematic process to follow: specifically applying for and receiving ethical clearance before moving to the data collection process, and the recruitment process. Further, how I accessed potential participants and negotiated gatekeepers will be discussed, followed by recruiting participants from my perspective, detailed in my reflective journal. Figure 4 illustrates the research action steps taken to collect the data, outlining the data collection stages.

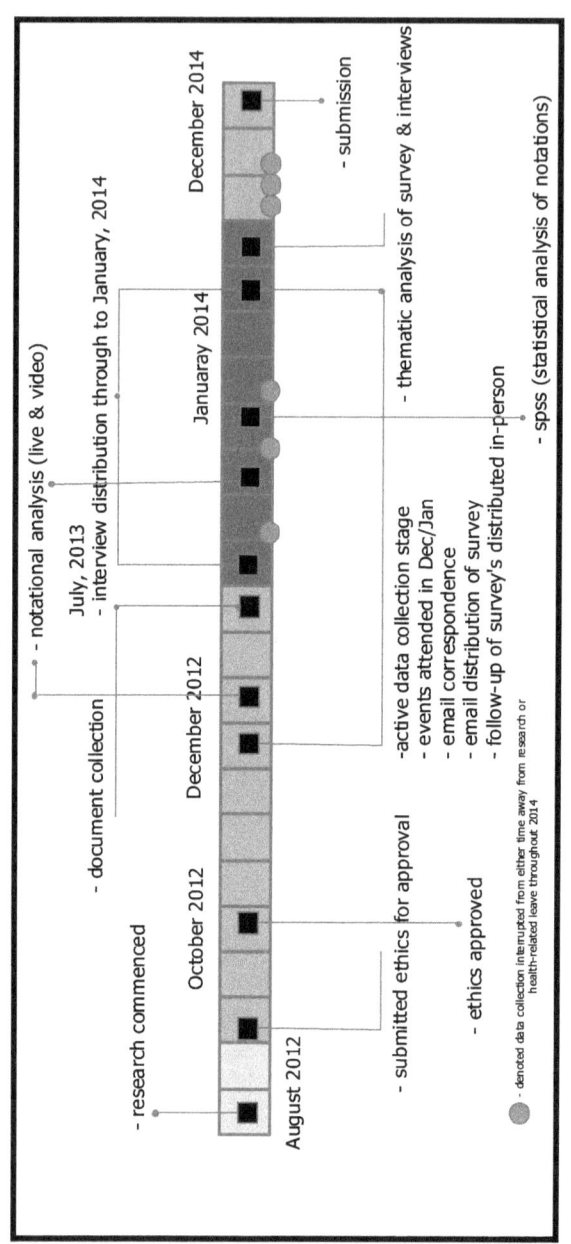

Figure 4 Research action plan illustrative of the steps taken during the data collection process and research journey.

Ethical clearance and ethical process

I sought permission from Griffith University's Research Ethics Committee at an early stage due to the pending opportunity to collect data in the Australian Summer of tennis. Ethical clearance was received in November 2012, as outlined in Appendix B. Permission was needed to initiate communication with potential participants (coaches) and to receive approval to distribute the survey and follow-up interview questions, ensuring that the survey and interview questions adhered to ethics policies and guidelines.

I ensured that all participants remained anonymous by de-identifying the data and not requiring the names of participants (although this was made optional if they were willing to complete the interview). I asked for permission to observe player performance in a public environment and record competitive matches to identify the use of the V Position. All recordings were noted as publicly available (for instance, open to the public to watch) and were used for pure observational purposes in identifying balance, height of point of contact, knee bend and shot outcome when viewing the forehand groundstroke. I also specified the use of de-identified data storage. Original data is securely stored on Griffith University password protected intranet, specifically in the LimeSurvey tool (survey design developer and distributor).

I requested informed consent from all participants, clearly outlined at the commencement of the survey and indicated on the first page of the LimeSurvey design. Participants provided their survey consent by submitting the anonymous survey online. Potential interview participants were recruited from this survey by adding the opportunity to provide identity and contact details at the end of the survey. These participants were made aware of the purpose and intent of the research through another informed consent form distributed at the time of communication (Appendix C).

In order to collect the required data and to access potential participants, I contacted the tournament organisers of the Brisbane International and the Australian Open via email to inform them of the intended research and to ask for their assistance with the objective of gaining the cooperation and approval of the selected tournaments. I was unable to gain the assistance of the tournament organisers, but it is important to indicate that I made them aware of my intentions and at all times throughout the in-person data collection process. I ensured I was wearing a Griffith University lanyard with my information in case anyone queried my research intentions. The de-identified collection and storage of survey data enabled me to respect the privacy legislation of my University, Australian laws, and international privacy laws of the countries of origin of participants, in accordance with Griffith University's ethics policy (http://www.griffith.edu.au/research/ research-services/research-ethics-integrity).

I required an ethics amendment towards the end of the data collection process to include participants who did not complete the survey (Appendix B). This was necessary to ensure potential participants for the interview questions received informed consent attachments through email correspondence. This amendment was necessary because of the low conversion of de-identified surveys into identified potential interviewees. I started contacting coaches directly via email and other social media outlets to increase this participant pool. This amendment was approved and a new direct interviewee recruitment process was initiated.

Data collection process

To begin, I sourced publicly available images of players when they were performing the theorised V Position. These images serve to illustrate that although I had theorised what the V Position might look like through photographs, my research objective was to identify the lower body positioning/performance characteristics of the V Position in practice.

Video footage of the players competing in the Sydney, Perth and Melbourne events was obtained from free to air broadcasting, televised by the Seven Network and Network TEN (primary Australian broadcasters), and recorded by Griffith University Network and Communication Services via MythTv. Over the years having viewed these tournaments before, I was aware of the quality and extensiveness of the coverage, which formed the video footage included in the observational analysis.

To distribute the survey, I approached coaches in person at the events I attended. It was foreseen that this may have been difficult due to appearing as a standard spectator with no endorsement from the tournament organisers. My objective was to make contact with as many coaches as possible, in between observing matches, and encourage them to complete the survey and interview questions. I did foresee that time was a major issue for participants whilst at tournaments, so additionally asked if they would prefer me to email them the direct link (for the survey), or if they did not have enough time, to complete the interview questions (requiring a smaller time commitment).

Participants who answered the interview questions provided additional data, at times furthering their answers to the survey; or alternatively, if they had not completed the survey, added another dimension for coaches who had completed the survey but did not wish to answer the interview questions: all participants would remain anonymous.

After a 12 month period of survey ($n = 46$) and interview ($n = 9$) data collection, I proceeded to analyse the participant responses through manual thematic analysis. The use of thematic analysis allowed identification of reoccurring themes in the survey and interview responses. In turn, I generated codes and sub-codes to form a conceptual map of the findings. I also used Leximancer as an indicator to check the reliability, validity and relevance of my codes and themes. A manual analysis enabled a greater depth of understanding of what the participants were saying and allowed me to use my insider perspective, actively inserting my ontological views into the research process; something that software could not do.

The recruitment process

I had to go through several stages before I could actually recruit from my designated sample. I will explain how I accessed participants and negotiated gatekeepers before moving to recruiting participants from my point of view, outlined in my reflective journal. Here I will discuss the barriers in place to collect my data, my experience when dealing with and attempting to recruit elite participants, the world's best tennis players and coaches.

Accessing potential participants and negotiating gatekeepers

My data collection plan was designed with sensitivity to the multiple gatekeepers operating within the arena of international tennis. According to Jupp (2006), a gatekeeper is a person and/or authority with a set discourse who has the power to resist approaches, for their own reasons, due to the hierarchical positioning that the person and/or authority holds over the information that the researcher would like to acquire. One of my prime gatekeepers was Tennis Australia (TA), Australia's leading tennis authority.

I initially approached TA to gauge their willingness to assist in the distribution of surveys and to provide access to tournaments to reduce costs and allow me a closer point for analysis (for instance, with a media pass that would allow a sideline vantage point). I was informed that due to my research not exclusively focusing on Australian tennis (restricted to TA endorsed players and coaches), rather than the entire coaching cohort (given that Australian coaches also coach oversees and not purely Australian players), my research did not directly align with TA policy. I was informed over multiple emails (Appendix D) that TA policy states that the research must benefit the respective tournaments (for example, the Australian Open) or be based solely on Australian tennis players and their respective coaches within the organisation (Appendix D). I also found that many coaches associated with TA declined to complete the survey due to their affiliation. Some TA coaches did complete the survey, possibly due to their acknowledgement of tennis being an international sport where players and coaches work with different nationalities. I was not willing to change my research focus to conform to TA policy and thereby gain their assistance as I felt strongly about the need for international participants, and the international significance of the research. My premise was to benefit the sport overall and not purely one nation. Additionally, TA does not have a large number of the world's best coaches, nor players associated with their system: they work overseas with players of different nationalities. My participant pool would have been restricted and I would not have been afforded the opportunity to observe the world's best players: merely a handful. A consequence of Australian tennis players not frequenting the top 10 in the world at this point in time, and with only a small number inside the top two hundred. For instance, <6 female and <11 male Australian tennis players were ranked inside the top 200 as of November, 2014.

As my own national tennis authority declined to help my research endeavours, I went straight to the top, and approached the WTA and the ATP tours, hopeful that with their assistance in the distribution of the surveys I would immediately be able to reach my target group. Due to both the WTA and ATP acting as representatives of the players and their respective coaches being privately hired by the playing team, my research had to benefit the tournaments directly to comply with their respective research policies (Appendix D). As my research is set to benefit coaches and players in the performance of the forehand groundstroke and ultimate development of an optimal coaching pedagogy, neither tours were able to be of assistance. I could have opted to alter my research project to get assistance from the bodies of authority, but I was adamant about answering the research problem with participants from all nationalities. I went to my last resort of governing tennis authorities: the ITF. To my dismay, they referred me back to the WTA and ATP tours for assistance. The responses I had received were problematic, in that Ethical Research guidelines and subsequent clearance from the University stressed that my research cannot directly benefit the participants, and these gatekeepers were unable to assist as a direct consequence of providing benefit. I had hit a road block.

As a coach and former player within the TA system, I had anticipated TA's knockback. As the WTA and ATP tours are privatised, found to be owned in part by the playing body (the players) and the entity (the tour), their information and resources would not be easily accessible and the tours themselves would not have permission to distribute personal contact information. At this stage, I conceptualised a multi-dimensional approach to access participants. To commence this approach I started in my own network of contacts as a professional coach, and allowed the snowball effect to take place. I used my personal social media outlets (primarily LinkedIn, Facebook and Twitter) in an attempt to contact participants. This method of accessing participants was a time-consuming process. LinkedIn acted as the best starting point and allowed me to apply snowball sampling methods: the more contacts I gained, the more contacts I had access to, but this was to dry up after only months. I had used up all my resources.

I had to think outside the box to access potential participants. I went ahead and contacted major national tennis federations that had an English contact (including North America, Canada, England and Sweden among others). Although my research was limited to English speaking countries, this did not stop me trying to contact other nationalities with the hope that English was their second language. This process was time-consuming and as a consequence of receiving similar replies as TA or not at all, I started the drawn-out process of searching for individual coaches located around the world who had the experience I was looking for, having coached or currently coaching a player ranked inside the top 200 on either the WTA or ATP tour. I started a country-specific search through Google, using different key words that would identify tennis coaches (for example, "tennis club Germany", "tennis academy France", "elite coach Canada", "high performance coach New Zealand", and so on). I made a list of countries, where English was spoken or where tennis was a well-known sport, and went through the first twenty pages of a Google search, sometimes more, until I could locate no more possibilities. This was a long and arduous process. I did not foresee that this component of my data collection would be this challenging, or time consuming, with the emails sent well into the hundreds. Ultimately, I either made contact with or attempted to (where there was an email available) with all coaches currently (at time of research) coaching players inside the top 200 on either the WTA or ATP tour. I also contacted every player inside the top 100 on either tour where contact details were available; I did not proceed after the 100 mark due to the very poor response rate I was receiving. This was at times a very disheartening process. However, eventually the desired number of participants was achieved.

Map Briefing: An Overview

The barrage of gatekeepers in *The Elite Research Method* borders on the absurd. It is unquestionable that the central premise of uncovering the extraordinary was near secretive. Those who came from outside the acceptable norm, for example, *non*-conforming scientists and/or *not* adhering to a set federations criterion, were barred – and strongly contested. If anything is amplified by *The Elite Research Method*, where there is a will – specifically aligned with first-hand experience, there is a way.

Invaluable in the research process is the role of experience and its *tie* to the overarching premise. In this case, experience was gauged by over a decade of immersion in elite sport as an athlete, with a direct transition, *overlapping* as a 'high performance' coach – having coached athletes on three *continents*. This experience in itself lead to the desire of *answering* the research question, having come across this divide mutually as an athlete, and as a coach.

This overlapping and personalised experience set the foundations for a strong undertone and grasp of the gatekeepers. Widely known in the literature is that great players *scarcely* make great coaches. However, it is also prevalent that those immersed in the sporting arena are *unaccepting* of the scientific literature. Furthermore, typical coaches inside these arenas are equally disapproving of the contributing sciences – particularly if the scientist isn't an *elite* themselves. However, if that coach is a genuine elite in their own right, denoted by their accomplishments in developing a 'top 10' player, this level of acceptance changes. Although, if that coach has *not* achieved this distinct level of success, the majority are equally prejudice and more likely to accept a coach as a former player, regardless of coaching success.

To simplify the breakdown, the majority of coaches across the sporting spectrum do not welcome external scientists. Whereby, external is denoted by not being a previous *internal* member. Internal is thus denoted by a previous elite player, thereby already accepted by the internal community, or coach that has already transitioned as a prior player inside this community. Therefore, a scientist that is external from this community, regardless of their prior experiences as a player and coach, remains external to the majority of those inside the sporting spectrum. The direct rational for this, as identified through the significant scope of *The Elite Research Method* was simple: the threat of the unknown.

Predominantly, people are comfortable, regardless of their set community, of external influences. A small portion of this community oppose this feeling and are a direct contrast – they welcome the external for what it affords them: *newness*. In relationship to *The Elite Research Method* this is afforded through scientific findings. The dilemma is then clear – with a significant portion of the community opposing the unknown, in particular if it is not from 'one of their own' then it becomes transparent as to why the gatekeepers are so vast, and wary.

To some, science is unwelcomed. More so, when a scientist seeks to enter a community as an external person – *with* experience, they are mutually in competition with the internal community members. Regardless if this is *not* the intention of the research premise, as with *The Elite Research Method*, internal members already have their minds made up due to this newness coming from an external person – highlighting the nature of this secretive stand-off – privy only to internal members. Or rather, this is what was hoped, but *The Elite Research Method* was used to unveil.

If experience was *not* present and if the research had *no* direct tie to the researcher, these blockades may not have been as vast. Yet, *without* the experience the level of understanding of the internal *favoritisms* would have been misguided. Subsequently, the notion to bypass the criterion of the gatekeepers may have flailed *without* this experience – amplified as an already external member of the spectrum in question.

Significance here is centered on experience. Without experience, the project premise risks being objectified by gatekeepers. This was experienced first-hand in the application of *The Elite Research Method* and why stringent processes were put in place to counteract these risks – furthered by prior *experiences*.

The immersion into a project premise is key – yet equally is the ability to step-back with caution and reflect on its overall progress. A key proponent to *The Elite Research Method* is the recall of the recruitment of participants – internal community members, and the scene where the data was collected. The narrative involved in this recount adheres to the qualitative *rulebook*, whilst also acknowledging its quantitative *mutual* denominator through a refined recollection.

Qualitative research encourages the use of a narrative voice, in direct contrast to quantitative research where there is *no* place for this reflection. In order to conform to the set *rulebooks*, whilst adhering to both disciplines – theoretical and statistical expectations, conformism was necessary. The end result was a culled and refined narrative that further encompassed the procedure of the notational analysis and the distribution of the survey.

Quantitative does have its advantages, whilst qualitative its disadvantages. The former did not rely on internal community members' participation – it could proceed irrespectively; the latter however did and thus was at the mercy of their participation. Recall, 'if anything is amplified by *The Elite Research Method*, where there is a will – specifically aligned with first-hand experience, there is a way.'

Recruiting participants: my reflective journal

What now follows is my experience in recruiting participants from my personal point of view. I begin with discussing the scene or environment that was my data collection playground, followed by how my notational analysis was conducted, and I will then discuss how the survey was distributed and the difficulties I faced. Following up potential participants to attain the desired numbers ($n = 55$) for my survey and interview methods will follow.

The scene: where I collected the data

My initial data collection had two objectives: to observe the best players in the world perform the forehand groundstroke, and initiating contact with some of the world's best coaches to get them to complete my survey, and if I was fortunate, also complete the interview.

There were two tournaments where I had the chance to collect this information, Brisbane and Melbourne, in a limited timeframe. I had less than 5 weeks to collect all the data needed. The benefit of attending these tournaments was the boundaries already in place; the nature of these tournaments meant that I knew I was going to be observing or talking with the best players in the world.

Recall that the participant group was limited to the top 200 players. This was a ranking range that I felt best demonstrated possible Grand Slam contenders. For instance, the top 104 players in the world receive direct entry on both the WTA and ATP tours at the Australian Open (Melbourne), whereas players who enter the qualifying draw can range anywhere up to and around 200 in the world rankings (sometimes higher, including wildcards). This cut-off ensured that the best players were considered (those ranked closer to number one in the world), whilst players who had the potential to be playing against this elite group were also considered. The Brisbane International had an even stricter direct-entry guide, making it even harder for the lesser ranked players (outside the top 100) to gain entry into the event.

The range of rankings (1-200) allowed me to look at the differences between performances at the elite level, for example the top 10 players, against those ranked outside this elite ranking range and to see if a difference existed in performance differences when observing the execution of the forehand groundstroke. Likewise, selecting the coaches to complete the survey and interview tools was streamlined by the ranking cut-off: they had to have coached or currently be coaching a player inside this ranking bracket (1-200).

The importance of the ranking range in place, for players and coaches, was to ensure I was observing and corresponding with the elite: the premise of my research question, as I wanted to find out what the elite were doing in regards to performing the forehand groundstroke and how coaches at this level were communicating these performance characteristics. A top 200 ranking range allowed me access to a potential sample size of 400: 200 on either the WTA or ATP tours. When accounting for coaches who no longer coached a top 200 player, this sample size grew as being able to talk to all top 100, let alone 200, coaches in the world would be difficult due to access to elites. Similarly, I knew the difficulties in observing all 100 players in the world.

Notational analysis procedure

The observation process was straight forward. My only challenge was getting the right seat to ensure I was able to get an appropriate seat to allow for a non-compromised viewing angle. As the Brisbane tournament was on a smaller scale, I was able to obtain a front row ground-level seat most of the time. Although the Melbourne tournament was extremely busy due to multiple matches occurring at the same time, I was able to view the less popular matches at times when I was unable to obtain an appropriate viewing angle for the more popular matches (featuring the top ranked players in the world).

I had organised my observation chart and so I was ready to commence my notational analysis from the beginning. What was not expected was the number of matches I sat through with less than 30 forehand groundstrokes notated. This made the process extremely time consuming, but on other occasions, I observed players hitting 30 forehand groundstrokes in 20 minutes to 3 hours, showing the diversity of the game and how different players have varied strengths.

In Brisbane I was able to observe approximately 30 elite players throughout their match, although the real number was slightly higher as some matches were excluded due to the criterion not being met (30 observed forehand groundstrokes). Centre court matches were notated from recorded footage at a later time. For the Australian Open, I was able to observe approximately 90 matches on mainly the outside courts. Players who were competing on the two main courts (Rod Laver Arena and Hisense Arena) were unable to be observed live due to financial limitations, but notations of their matches were still made via recorded footage.

Through the observation tool I was able to control the number of matches that were observed and ensure the players were included in the criteria. When deciding which matches I was going to observe on any given day, I would look at the draw the night before (when the draw and match times were released). This process allowed me to be somewhat organised and know which matches were the best to be observed (to avoid double ups and ensure only elite matches were recorded). At the end of my five-week period spent observing elite players, approximately 120 players had been observed. I could have increased this number; however, due to the limited number of seats available in the second week of the Australian Open, viewing angles would have been compromised and I did not want to affect the quality of already collected data (see Appendix E for the Work Log of the data collection phase).

Distributing the survey

The structure of tournaments allowed for a given end-date (five weeks). The distribution of the survey, however, relied not on how many I could distribute, but how many replies I could manage. My initial distribution of the survey commenced at the Brisbane International in January 2013. The Brisbane tournament is one of the tier lead-up tournaments to the Australian Open, where I had eight days (including qualifying rounds) to approach potential participants and ask if they would participate and complete the survey; I would then repeat this a week later in Melbourne over a fourteen-day period (including qualifying) and ask coaches to complete the survey within a six-week timeframe.

At first, I was very hesitant to approach participants, because I had admired some of them, as a player (the players), and as a coach (the coaches). Both Brisbane and Melbourne attract the best players in the world with rankings inside the top 200 on either the WTA or ATP tours. I not only had the opportunity to observe some of these world-class players, but also to talk to their coaches. This was a juggling act as I had a limited timeframe to observe matches, whilst also approaching coaches in-person to complete my survey. Therefore, I had to multitask: observe one match, approach one coach, and at times I would approach coaches during a change of ends if I had the time. But this was very hard: walking up to some of the best coaches in the world was a daunting task. Due to the eliteness of this group, I made sure I had guidelines in place, particularly due to my research intentions.

In order to distribute the survey, I formed a set of guidelines to follow, ensuring that utmost professionalism was maintained and that I was able to separate myself from the crowd as a researcher opposed to a fan:

- The player/coach was to be by themselves or with a fellow player/coach;
- A player/coach was not to be interrupted if they were walking hastily away from their origin;
- A player/coach was not to be interrupted during practice, however a coach was able to be interrupted during practice if they were near the sideline and were not actively coaching the player;
- A court-side person who appears to be a part of the coaching team was able to be interrupted providing that they were not actively busy or assisting the coach with the training session;
- A player/coach was permitted to be interrupted if walking on or off court prior to or after practice (they were not to be interrupted if walking on or off court from a match).

I followed these guidelines so my data collection process was consistent and professional when distributing the survey and observing matches. A purpose-made business card was also tied to these guidelines as it was handed out to each person with survey links attached to the back of the primary card, with the secondary card (attached) identifying myself as a researcher with specific contact details to the School of Rehabilitation Sciences (later, Allied Health).

The manner in which I introduced myself to potential participants and asked if they would agree to completing the survey took several iterations and was adapted over the days/weeks due to response rate and any language barriers I felt were present. Initially I started with: "Excuse me, I'm doing research on elite tennis players and coaches, if you wouldn't mind completing this survey when you have time, I'd really appreciate it, thank you." I thought it was best to be quick and to the point whilst trying to ensure that I was specific. I also ensured that I did not walk away until they had asked any question they may have had. Some potential participants took the business card and replied with "thank you" without asking any further questions.

I did find that my initial speech was too long: I was being interrupted. I decided that I had to come up with an even shorter phrase that would best represent my research. I also had to rely on that if the participant had any questions, these would be answered on the business card or if they wished to contact me this was also made clear on the business card. I changed my initial speech to: "Excuse me, may I?" Then proceeded to hand the potential participant the business card and briefly explain what I was doing if given the opportunity, that is if the potential participant did not walk away.

Over a five-week period distributing the survey, I experienced 10 people who ignored me and walked away; a handful more who were discouraging and put me off approaching more coaches for a little while, before I plucked up the courage to continue to approach more potential participants. I had the opportunity to approach >100 potential participants and pass on my survey to complete. These potential participants however were not all English speaking; although I tried hard to articulate what my survey was about, this proved to be difficult due to language barriers. Also, the potential participants I did approach were not guaranteed to be inside my ranking range, so I ensured a disclaimer was included at the start of the survey that only coaches who fit this criterion to complete the survey. In the instances where I contacted coaches directly (not via the survey), I ensured that this disclaimer was either present or clearly explained.

I came to realise that my introduction to these potential participants might be an issue. It was suggested that I start by introducing myself and letting them know I'm doing my PhD. I had thought of this earlier, but feared that I already had little time to speak with them, and lengthening what I wanted to say to the coaches and players (or team) was a little daunting. I took the advice and changed the introduction:

> Excuse me, hi my name is Ashley and I'm doing research on elite tennis players and coaches; if you wouldn't mind completing the survey in the next month or so [this changed to after the tournament at times] along with your coach/player, I'd really appreciate it and it will contribute towards my research, thank you.

I should note here that originally I was after the player's point of view, though this ceased when I realised the difficultly in recruiting players. It was near-impossible to recruit this group of elite players, particularly when I was having enough difficulty recruiting coaches.

As soon as I introduced myself, surprisingly, they wondered who Ashley was and found that if the person did not wish to talk to me, it wasn't offensive to my research as they didn't even know I was researching, as I was only able to get out,

"Excuse me, my name is Ashley," before they either said "My English is no good," "I can't understand," then turn away or say, "No I'm busy". This new introductory phrase was not necessarily a direct effect on the research itself, but the person not willing to take the time. However, it was apparent that by introducing myself it was clear that I was not a fan wanting a signature; a status that I was very weary of and if I felt as though I had put myself in that situation I would not proceed with the introduction. When short on time, which was occasionally the case when a coach or player started to walk away, the most important part of the phrasing was to ensure key words were used earlier and thus changed my introduction to:

> Excuse me, my name is Ashley and I'm doing research on players on the ATP and WTA tour for my research; if you're able to complete this survey in the next six weeks, I'd really appreciate it: thank you.

I used this speech when time was short, but also found that the acronyms ATP and WTA gained the coach or player's attention, due to them being affiliated with the respective tour. It was also important to mention the word research, as it was found early on that no one recognised the term PhD or doctorate (after testing them all); the safer option was 'research' which I thought to be a universal acronym for research, regardless its calibre. When given the chance, the person who heard the word research was more understanding and felt less imposed on. I learned that these terms were important when talking with people from around the world with limited time. It should be noted here that unless the player/coach was well known or I personally recognised them, I was unable to identify their ranking and/or nationality, or if they spoke English. Not knowing the players' or coaches' player rankings before I approached them was not of concern due to the cut-offs for this Grand Slam tournament, whether or not the person was a coach or player (or someone else) and therefore a potential recruit was difficult to recognise at the commencement of each tournament. If I was unsure if they were a coach or player, prior to introducing myself I would ask, "excuse me, do you have a player in the tournament," or "excuse me, are you playing in the tournament," and depending on their answer, I either proceeded to introduce myself or thanked them for their time.

Through my interaction with coaches and coaching teams, it was found that there were some who seemed as though they may be willing to help me distribute the survey. When I felt this was the case I would ask, "if you have any colleagues who wouldn't mind completing the survey that would be very much appreciated," depending on the circumstance, and I would give them an additional few business cards. Overall, I distributed 150 business cards to 90 coaches, players, family members, hitting partners, or physical trainers.

The follow-up: waiting to hear back from participants

Due to the time period specified in completing the survey, it was a waiting game of two weeks post tournament to see if the business cards were an effective means of recruiting participants. I was to find after the six-week period had passed that my distribution of the survey was not effective, with less than five people completing the survey after three months. Although this number did triple a month later, I was disappointed in the overall result. I speculated whether having the option of a hard copy survey printed out and a pre-paid envelope for a prompt return may have recruited more participants as opposed to having to go online. This process would have increased costs and perhaps would not have been as efficient when I only had a matter of seconds communicating with potential participants. In hindsight, the initial timeframe should have been narrower, although February 2013 was host to Davis Cup and Federation Cup competitions (where athletes compete for their respective country against other countries in their group). I did not consider the implications of this timing.

After five months had passed since my initial distribution, it was clear that I was not generating enough participants. I did have four contacts acquired over the time, and who I could ask if they had had time to complete the survey. Due to the anonymity of the survey, I was unable to identify who had completed it and who I could ask again. At the end of the five months, 14 individuals had completed the survey. The survey design presented a challenge, as due to the layout of the survey I was unable to confirm what stage of the survey had been completed: 5 registered as complete, and 10 listed as incomplete. This indicated that not all of the survey questions had been answered. I understood this was a start and that the participants were able to come back and finish filling out their answers.

Throughout this process, I made contact with numerous coaches who did not meet the criteria. The survey tool made it clear, through the information sheet (Appendix C), that coaches who did not fit the criteria should not proceed to complete the survey. The survey had a qualifying question, however, that when the participant entered the response on current player's rankings (whom they were currently coaching, or had previously coached), it was clear that this ranking range was exclusively within the range of 1–200. Although I was unable to stop these coaches from entering the survey in the first place; as a result some surveys were only partially answered.

In order to reach a total of 14 participants, I had spent a week per month over the past four months distributing the survey. This predominantly included sourcing contact details and sending out emails asking potential participants to complete the survey. As of June 2013 I had searched 90 % of countries (if not more) and their respective local and national tennis academies, in hope of finding suitable participants to complete the survey.

After an additional five months had passed, the survey completion had grown significantly. I had spent the following five months largely focused on distributing the survey and also following up with participants who left their email address at the end of the survey (where it was made available). The total number of participants who had completed the survey jumped to 46 – 18 complete and 28 incomplete. The incomplete margin varied from only one or two answers completed through to almost all questions answered except for a select few. At this point in time, I had had the survey live for nearly 12 months, and within this timeframe although I was yet to generate the quantity of responses I had hoped for, the data that I collected began to tell a story I had not predicted. Over this time, I had been able to commence the distribution of the interview (to participants who provided their contact details), and a total of 7 responses had been finalised with 8 pending (that is, the participants had received the interview questions and I was waiting for their completion with regular follow-up reminders).

I decided to cease recruitment after the 2014 Brisbane International (January), which allowed me to make contact with an additional 9 elite coaches. I then decided to close the survey in February 2014, giving potential participants a month to complete the survey. Of the 9 participants who were contacted, 1 participant returned the interview questions to bring the cumulative total of interviews to 9; although I was unable to determine if this participant also completed the surveys. My survey total had reached 19 complete and 37 incomplete. Incomplete surveys increased the number of overall responses to 46, with varied questions ranging from 19 responses to 72, highlighting that the survey did in fact receive a large number of views and responses, even though less than 50 % of participants completed the full survey. The high number of responses was generated through multiple choice questions, where I allowed participants to select more than one answer. Ultimately after nearly a year and using a variety of social media outlets and sourcing contact details via a rigorous online search, I was able to generate a total of 46 surveys and 9 interviews with data closure at the end of February 2014.

Map Briefing: An Overview

A simplified and brief recount of the reflective journal afforded a personalised response, yet its drawback is in its conformism and incomplete tale. Nevertheless, the *rulebooks* were adhered to in *The Elie Research Method* and the narrative is merely one 'give' for the next 'take' in merging methodologies in the *real-world*.

The *framework applied in the real-world* is a cohesion of science – theoretical and statistical with an overarching premise. No one knows at the beginning of a project its answers – after all, it's the answers that are sought in order to comprise the secondary questions earlier discussed. The *real-world* affords science to come to life – the witnessing of data in reality and how it shapes our world.

Irrespective of the data sought, its *real-world* application is necessary to address its theoretical stance, whilst its statistical inference affords scope for scale. The foundations of *The Elite Research Method* is courtesy of both stance and scale – without either the *methodological framework* would not have its connotations unveiled.

A direct result of the coupling of stance and scale is the analysis – theoretical *and* statistical, respectively. *Frameworks* require data analysis steps for not only structure, but in direct relation to its *map*. The *framework* applied in *The Elite Research Method* ties back to its initial secondary questions that in turn points toward answering the overarching primary question. The same applies for the project premise at stake – the analysis steps provide the pathway to answering its initial premise. Furthermore, both stance and scale are of focus – through individual analysis.

Scale is addressed through the undertaking of notational analysis, whilst stance is the output of sieving through both survey and interview data. Due to the respective *rulebooks*, the analysis must adhere to the set processes in place. However, collaboratively the mixed methods design affords a new output in its application and use in *elite research.*

In order to divulge the collective response to the overarching premise, qualitative and quantitative methods of inquiry are key – separately, to allow the data to unfold naturally. Only then can a collective analysis move forward in its assertion. The fundamental role of *The Elite Research Method* is its *methodological framework* and its transitional application, providing a map to guide the researcher or project lead. The use of analysis inside this *framework* is critical to its disclosure.

Data analysis steps

The data collection process followed three distinct methods: observation, survey and interview. In order to record the observations, notational analysis was used, and to report on the survey and interview results, thematic analysis was employed. The analysis steps will be discussed in the following sections.

Notational analysis

Each individual variable was recorded against the criteria for each individual match. Each observation chart was systematically studied separately to generate individual numbers that represent the player's performance when contacting their forehand groundstroke. This step required examination of >150 matches, whilst recording each individual variable, including matches that fell short of the 30 forehand groundstrokes. All matches were recorded in case there was a later benefit that might appear in the matches with fewer than 30 shots recorded. The total number of individual variables that were recorded for the criteria and transferred into SPSS totalled 11,070 notations to generate the 121,770 notation relationships, used to identify the criteria and combination of variables (mixed criteria). This was a very time consuming process, but necessary in order to identify the frequency of the V Position amongst the world's elite players.

As discussed earlier, to ensure the reliability of the observation tool and notational analysis, inter-rater reliability measures were employed. In order to perform the inter-rater reliability testing, another rater (an experienced research assistant) was engaged to observe and record independently. The additional rater recorded a total of five complete matches alongside the primary researcher without consultation of results at the time of scoring, which produced an additional five completed observation charts. These charts allowed the inter-rater reliability tests to be calculated. This data set included previously notated matches from recorded footage that were notated a second time in conjunction with the additional researcher to ensure the inter-rater reliability results were based on the same data set used by the primary researcher. Being post-data collection period, all inter-rater reliability measures were generated from recorded footage only.

Thus, the overall findings, consisting of 121,770 individual notation relationships, and 11, 070 notations across 123 players, employed inter-rater reliability measures. The inter-rater reliability was reported by using an additional rater to use and report on five matches that were also observed by the primary researcher (author) to report on the reliability of the observation tool, as discussed in the researches statistical inferences section. The extensiveness of the calculated variables, generated charts and tables, and the tests performed (reliability, correlations, *t*-tests and descriptive statistics) present a comprehensive representation of the lower body movement characteristics of player participants in this investigation, giving confidence in the overall findings of the application of the V Position by elite tennis players.

Survey and interview analysis

In order to analyse the survey and interview questions, each participant response was initially manually coded. The generated survey results were extensively analysed, having been automatically produced through LimeSurvey, for each individual question and the subsequent response and/or selection. Each closed-ended question provided multiple choice selections. However, the format of the survey included an open-ended question format, encouraging the participants to further explain their answer/selection to provide additional data to be incorporated in the findings, providing further scope for analysis. As a result, more data were made available to work with than originally thought, contributing towards the depth of the survey and interview data. All of the survey data were used to generate subsequent charts and graphs to illustrate the overall responses. Open-ended questions embedded in the survey and interview responses were analysed and common themes in the data were manually identified and coded. After manual coding, software-based coding in the form of Leximancer was also conducted.

The survey and interviews were used to gather data on how the participants communicated the performance characteristics of the forehand groundstroke, and how their knowledge of these characteristics influenced their coaching pedagogy. Common themes were identified from the interview responses that were also found through the survey results. These themes were then analysed to begin the process of answering the research questions. These themes provide the framework for the findings of the survey and interview questions.

Research Conclusion

Asserted by Alvesson and Skoldberg (2009), research methods are complex in nature, and in order to accommodate the premise of this research, a mixed methods approach has been designed and implemented. The interview method permitted access to elite participants, discussed by Moyser and Wagstaffe (1987) as typical and pragmatic when accessing expert knowledge. The notational analysis, a form of data collection, and suggested by Dogramaci et al. (2011) as producing high quality data irrespective of the subjective role of the researcher, was used to record the elite participants against the designed observation tool and its criteria. The survey tool, discussed by Long (2007) as an effective way to collect data from geographically dispersed participants in a timely manner, was also utilised, developed and deployed.

Snowball sampling and data triangulation assisted in the data collection process, whilst the notational analysis technique and the reliability methods applied were used to support the validity of statistical findings. The observation tool formed an important part of the research methods and research action steps, acting as the key data collection method to generate the statistics on the V Position's criteria and its variables. The research action steps detailed the data collection process and methods of analysis for the notational analysis, survey and interview findings.

The objective of the research process was to generate ample data to identify the performance characteristics of V Position and its application amongst the world's best players. The statistical inferences reporting presents the findings from the notational analysis of elite players' application of the V Position, particularly noting individual and combined means, frequencies and significance of individual criteria and thus characteristics of elite player performance in the forehand groundstroke. The frequency of the V Position according to player ranking will also be established.

The thematic analysis of the findings incorporates the collective schematic and its results. Overall, the data collected conforms to the collective process shared in Figure 3 and centres on answering the research questions. A narrative paradigm was employed to outline the direct association of the V Position, its communication, and how it assists in the development of an elite coaching pedagogy throughout coach education practices.

Map Conclusion

The design of *The Elite Research Method* required an encompassing schematic that afforded versality in its interpretation for its transitional use. Science itself has a reputation to be hidden away and accessible to *only* scientists or those who are well versed in the respective fields of focus. Too often an outsider will greet the sciences with open arms only to be turned away by its jargon and non-applicability for its transition into the *real-world*. After all, this is personal.

All researchers and scientists alike were once external to their field. Too often, the *real-world* escapes the human immersion into the sciences and its ultimate applicability. But, recall the emphasis on experience. For it is in fact experience that guides us and our use – whether internally involved, or externally. Experience affords perspective and the capacity to shape the environment of immersion to its *real-world* application. However, this shaping is mutually reliant on experience and depth of knowledge of the interval verses the external spectrum.

The role of the respective spectrums in *The Elite Research Method* referenced their corresponding disciplines brought together through cohesion – the shared *framework*. However, health *and* education not only brought to life *The Elite Research Method*, their respective *rulebooks* set at holding each afar from the other was used consciously to form a schematic. In turn, this schematic used experience as an inadvertent measure to counter the unhinged relationships in the opposing scientific fields that ultimately brought the spectrums closer.

Practicality is key – in science and in life. Although not as commonplace as it may be suggested. The role of science in our world is insurmountable – it provides scope for advancements, whether theoretically or statistically speaking. Significance is placed on the practicality for application. For although there is significant scope, well-known, for the sciences internally – there is limited scope for the sciences externally: their *real-world* applications.

An emergence is shared. *The Elite Research Method* would not be possible without the role of experience that was inadvertently used in its unveiling. Yet, both internal and external experience – scientifically, and prior personal experiences, contributed to its output. Although careful consideration has been taken to null the complexities presented within, there is no mistaking the sciences – complexities run rife, yet it is the explanatory nature that affords its transitional use.

Qualitative research by far affords *real-world* transitions through the use of its narrative. However, it is equally arguable that although a story is unfolded, the thematical analysis ensued is rife with its complexities. The same is apparent for quantitative research. Numbers are a common notion for the depiction of a result – yet their statistical inference is likened to the aforementioned analysis – complex in its nature and hard to transition into the *real-world* – externally speaking. This is the downfall of scientific research – understood by its internal community, yet hard to comprehend for external community members.

Recall the gatekeepers that threatened the absolute foundations of the overarching premise. If it was not for experience, these gatekeepers may very well have succeeded. Yet, scope for the internal and external workings of the project and the respective community members afforded these risks to be countered. Thus, the true hero of *The Elite Research Method* is experience.

As is with life. The *framework* shared is a map in itself for the undertakings of *elite research*. Mutually, the *framework* depicts the *bird verses cat* relationship and the synchronicity afforded through acceptance. Recall the beginning whereby the disciplines: applied sciences, simplified by health, and pedagogical research, simplified by education, have in turn their own *methodological* approaches – 'set on answering a question, whilst the other is drawn from the answer'. Without prior scope of experience, the *bird verses cat* would be at one's demise. Typical is the case in the sciences. But as shared, it doesn't have to be.

Cohesion in science affords the extraordinary to be unearthed, as *The Elite Research Method* was in unveiling the Optimal Performance Theory and The V by Dr. B©, herein shared as the V Position (Berge, 2017). But cohesion begins with acceptance of science overall – after all, scientific findings *are* factual evidence *opposed* to hearsay or opinion. And this is the dilemma faced when applying science to the *real-world* – lack of acceptance by the external community members.

To instigate a change in perception of the sciences and their application to the *real-world*, scientists internally must begin to address their respective use-cases to transition into the *real-world*. Without this approach, the sciences will remain closed off from external communities who are unable to gauge the *real-world* implications themselves. Mutually, negating *real-world* transitions amplifies the role of experience and its emphasis, or lack thereof and consideration in the overall research premise. This signifies the importance of multidisciplinary research – as amplified with health *and* education, and its increasing use in scientific practice.

Yet, to negate hearsay and opinion, it is the *real-world* use of *The Elite Research Method* that begins to quiet the misinformed – both internal and external communities. The division between the respective sciences and the ill-acceptance within – to instigate acceptance, an increased understanding of demand for the rigor involved in scientific findings and their resultant implications for the world as it is known, is fundamental. This begins with a science overlay of the map (Figure 3) that guided *The Elite Research Method* and its transparent consideration for scientific research for transition into the *real-world*.

Albeit science is vast in its respective complexities, it is the implications that afford *real-world* use. The complexities are representative of the scientific inferences at large and the language used to relay profound depth. There is no other field quite like it – after all, it is the sciences that broaden the scope of our world. Yet, in order to relay complexities to external members, *real-world* consideration is key. This task in itself requires immersion.

Recall, seldom does anything work by itself. Equally, consideration for the external *and* internal delivers on *two is better than one*. Likewise, one discipline is standard practice, a *cohesion* between disciplines broadens its *real-world* applicability. Individually, theoretical *and* statistical analysis afford inferences in their own domain – yet, consideration for stance *and* scale broaden the horizon of applicability.

The Elite Research Method is a scientific implication itself – the analogy of the *cat verses the bird* is analogous with the divisive relationships throughout. The *real-world* rings true – relationships take two. Harmony exists through acceptance – reliant on a broader scope of view. Thus, transitional science is a relationship – inclusive of immersion in the *real-world* and the inference of experience as to life. If communities are capable of extending beyond their own – internal to external, as external is to internal, the *framework* herein is boundless in its pursuits.

What Now?

Visit https://AM8International.com and https://Topicthread.com for what's happening now and the latest resources available.

Want to get in touch? You can find Dr. B on Topicthread @Dr.B where she shares the latest news from Topicthread as well as her current running endeavours, publications, and performances.

Books by Dr. B

The Secrets to Optimal Performance Success
(March, 2016)

The Secrets to Optimal Wellbeing
(September, 2016)

The Science of Elite Performance
(March, 2017)

The Secrets to Optimal Coaching Success
(March, 2018)

The Elite Research Method
(October, 2018)

What is Your Game Missing?
(March, 2019)

What is Your Game Missing, Now?
(October, 2019)

What is Your Game Missing, to Win?
(July, 2020)

I am Your Tennis Coaching Guru
(December, 2020)

The 7 Keys to Optimise Your Life
(December, 2021)

Appendix A: Survey

Section A: Questions 1 through 5

A1. Please indicate your nationality:

A2. How many years have you been coaching:

- 1-5 years ☐
- 6-10 years ☐
- 11-15 years ☐
- 15-20 years ☐
- 21-25 years ☐
- 26-30 years ☐
- 31-35 years ☐
- 36-40 years ☐
- 41-45 years ☐
- 46-50 years ☐
- More than 50 years ☐
- Other ▼

Please specify

A3. What is the ranking range of the player your are currently coaching:

- [ ] 1-5
- [ ] 6-10
- [ ] 11-15
- [ ] 16-20
- [ ] 21-25
- [ ] 26-30
- [ ] 31-35
- [ ] 36-40
- [ ] 41-45
- [ ] 46-50
- [ ] 51-55
- [ ] 56-60
- [ ] 61-65
- [ ] 66-70
- [ ] 71-75
- [ ] 76-80
- [ ] 81-85
- [ ] 86-90
- [ ] 91-95
- [ ] 96-100
- [ ] 101-105
- [ ] 106-110
- [ ] 111-115
- [ ] 116-120
- [ ] 121-125
- [ ] 126-130
- [ ] 131-135
- [ ] 136-140
- [ ] 141-145
- [ ] 146-150

Section B: Questions 6 through 10

B1. What was the ranking at the time, of the previous players (last 5) you have coached?

Player 1
Player 2
Player 3
Player 4
Player 5

B2.	Where does your coaching pedagogy (philosophy) come from?

- Other coaches I admire ☐
- My coach when I was a junior ☐
- My parents ☐
- Family members ☐
- Coaches I have spoken with over the years ☐
- My morals and principles ☐
- Other ☐
- Other ▼

Please specify
[   ]

B3.	What forms your coaching pedagogy (philosophy)?

- My morals ☐
- My beliefs ☐
- Coaching Games for Understanding ☐
- Game Sense Model ☐
- Classical coaching approach/theories (please explain) ☐
- Modernised coaching approach/theories (please explain) ☐
- All of the above ☐
- None of the above ☐
- Other ☐
- Other ▼

Please specify
[   ]

B4. **How do you communicate lower body movement to your players for the groundstroke?**

- [ ] Get Low
- [ ] Bend
- [ ] Bend your knees
- [ ] Lean forward
- [ ] Lower yourself to the ground
- [ ] Drop
- [ ] Leg Drive
- [ ] Drive through your legs
- [ ] None of the above (please specify what you would say)
- [ ] All of the above
- [ ] Other (please specify what you would say)
- ▼ Other

Please specify

B5. **Can you give an example of when you would communicate lower body movement to your player?**

B6. **Is a bend at the knees during the goeundstroke necessary?**

- [ ] Yes
- [ ] No

B7. Please explain:

B8. Do you believe players should bend their knees during the groundstroke?

Yes ☐
No ☐

B9. If so, at what point: A) Prior to impact and release (straighten knees) prior to making contact with the ball B) Prior to and during impact (straighten knees after the backswing is complete)

A ☐
B ☐
Other ▼

Please specify

B10.	Please explain your answer:

Section C: Questions 11 through 15

C1. Where do you coach your player to hit their groundstroke:

Above the hip ☐
Hip height ☐
Below the hip ☐
Other ▼

Please specify

C2.	Please explain your answer:

C3. How do you communicate the generation of power, for the groundstroke, to your player (what do you ask/tell them to do)?

- Explode through your legs ☐
- Explode through your upper body ☐
- Attack ☐
- Drive through your body ☐
- Push against the ball ☐
- Fasten your swing ☐
- Increase the speed of your swing ☐
- Use your upper body ☐
- Use your lower body ☐
- Stay low ☐
- Stay upright ☐
- Other (please give examples) ☐
- Other ▼

Please specify

C4. Please explain your answer:

C5. Have you heard of knee bend in the groundstrokes?

☐ Yes
☐ No

C6. Is it applicable?

☐ Yes
☐ No

C7. When is it necessary or why isn't it necessary?

C8.	Should a groundstroke be contacted at the same point (in regards to hip height) irrespective of pace?

Yes ☐

No ☐

C9. If different, why?

C10. Please identify the stance that is most appropriate for your player during their groundstroke:

low to the ground with knee bend (upper body leaning forward) ☐

mid knee bend, chest upright (lower body bent slightly, upper body straight) ☐

standing upright, feet apart (neutral position) ☐

Other ▼

Please specify

C11.	Please explain why you prefer this stance:

Section D: Questions 16 through 20

D1. How do you know when to tell your player to bend their knees (if at all)?

- When there is no knee bend evident ☐
- I don't have to tell him/her - they always bend their knees ☐
- I don't have to tell him/her - I don't encourage them bend their knees ☐
- When they are only slightly bending their knees ☐
- When they are contacting at hip height ☐
- When they are contacting below hip hieght ☐
- When they are contacting above hip height ☐
- All of the above ☐
- None of the above ☐
- More than one of the above (please specify) ☐
- Other (please specify) ☐
- Other ▼

Please specify

D2.	How do you know when your player should hit the ball at hip height (if at all)?

It's the most ideal position for the best results ☐

They don't ☐

For increased power ☐

For increased accuracy ☐

For increaed control ☐

None of the above ☐

More than one of the above (please specify) ☐

Other (please specify) ☐

Other ▼

Please specify

D3. Why do you encourage the player to hit the ball at hip height or why don't you encourage the player to hit the ball at hip height?

D4. Is there a difference, if any, between players who bend their knees at contact and make contact around hip height, compared to players who don't bend their knees and hit above (or below) their hips?

Yes ☐

No ☐

D5. Please explain your answer:

D6. Is there a coaching discrepancy (difference) when players should and shouldn't bend their knees?

Yes ☐
No ☐

D7. Is there a coaching discrepency (difference) where players should make contact with their groundstroke (in relation to hip height)?

Yes ☐
No ☐

D8. Please explain why their is a discrepency for: A) Knee bend and B) Contact Point in the groundstroke (if relevant)

A
B

D9. Statement: Players should make contact with the ball during the groundstroke, with knee bend whilst making contact at hip height.

Strongly Agree ☐
Agree ☐
Neutral ☐
Disagree ☐
Strongly Disagree ☐

D10.	Please explain your answer:

D11. What do you say to your athlete to improve their groundstroke in respect to lower body movement?

- [ ] Get low
- [ ] Lower yourself to the ground
- [ ] Drop
- [ ] Drive through your legs
- [ ] Stay low
- [ ] Bend your knees
- [ ] Drive
- [ ] All of the above
- [ ] None of the above (please specify)
- [ ] Other (please specify)
- [ ] Other ▼

Please specify

D12.	What do you say to your athlete to improve their groundstroke in respect to point of contact?

- Contact below your hips ☐
- Contact above your hips ☐
- Contact at/around your hips ☐
- Move to the ball ☐
- Move away from the ball ☐
- Make contact when the ball drops ☐
- Make contact when the ball rises ☐
- All of the above ☐
- None of the above (please specify) ☐
- Other (please specify) ☐
- Other ▼

Please specify

D13. **Please explain your answer:**

Section E: Questions 21 through 25

E1. Does your player understand what you say and take it in the correct context?

YES ☐

NO ☐

He/She understand but doesnt take it in the intended context ☐

He/She doesn't understand but takes it in the intended context ☐

E2. How can you tell if he/she understands?

He/She tells me they do ☐

He/She does what I ask so I can see they understand ☐

He/She repeats back to me to show they understand (paraphrase) ☐

He/She practices what was discussed so I can see they understand ☐

Other (please specify) ☐

E3. If you do not mind being contacted in relation to your answers in this survey and potential follow-up questions directly related to this survey, please provide your name and email address.

Name: ☐

E-mail: ☐

Appendix B: Ethical Clearance

GRIFFITH UNIVERSITY HUMAN RESEARCH ETHICS COMMITTEE

21-Nov-2012

Dear [researcher]

I write further to your application for ethical clearance for your project "NR: How the V Position is perceived, communicated and applied to the forehand groundstroke by coaches and players internationally." (GU Ref No: PES/45/12/HREC). This project has been considered by Human expedited review 1.

The Chair resolved to grant this project provisional ethical clearance, subject to your response to the following matters:

This application has been reviewed administratively by the Office for Research via a mechanism for research that has been assessed as involving no more than negligible risk.

Please note that in the case of student research, a member of the student's supervisory team must be listed as the contact person. The research student cannot be the listed contact person. Please also ensure that all materials identify the PhD Candidate as "student researcher", and ensure that all materials clearly state that you are conducting the research in your capacity as a Griffith University student.

The response to question E2 in the Checklist cannot discuss only the expertise of a research student; it must also outline the supervisory team. University policy is to regard the principal supervisor as the Chief Investigator and the student as a member of the research team. This is the case, even when the student holds an academic appointment. Applications must outline the relevant expertise of the supervisory team and how the involvement of the supervisor(s) will ensure the ethical, appropriate and successful conduct of this research.

It is recommended that you contact the various tournaments/venues where you intend to recruit potential coach and athlete participants and advise them of your research project, and seek their consent to approach the coaches/athletes at their venue and/or to post the recruitment flyer.

Please provide further information about the observational component of your research project. In particular, please clarify whether you intend to videotape athletes and/or coaches. Please also clarify whether you intend to observe only those coaches and athletes who have consented to participate in the interview or questionnaire or whether you intend to observe other coaches and athletes as well. If you intend to videotape athletes or coaches, you must provide them with prior notice that they will be recorded and seek their consent to the recording. You must also outline the retention, access, and destruction of this data. Furthermore, we note that it may not be possible for you to take video footage at tournaments or venues as they may expressly prohibit videotaping under their entry conditions. You will need to consider this for each venue you attend.

Similarly, it is recommended that you familiarise yourself with any privacy legislation or recording laws in other countries. Please ensure that you have also referred to Booklet 39 – Research Conducted in Other Jurisdictions.

It is recommended that the section on the Participant Information Sheet "Why is the research being conducted" be more succinctly set out (see section 7.4 of Booklet 22).

For student researchers, it is assumed that supervisors may need to access the data and this should be discussed in the informed consent materials.

Please correct the contact details of the Manager, Research Ethics.

The contact officer signing sF1 of the hard copy of the Expedited Ethical Review Checklist. If you did not generate a hard copy when you first submitted your application we can email a PDF copy to you.
The primary supervisor signing sF1A of the hard copy of the Expedited Ethical Review Checklist. If you did not generate a hard copy when you first submitted your application we can email a PDF copy to you.

An appropriate authorising officer, who is not a member of the research team, completing and signing sF2 of the hard copy of the Expedited Ethical Review Checklist. If you did not generate a hard copy when you first submitted your application we can email a PDF copy to you.

This decision was made on 21-Nov-12. Your response to these matters will be considered by Office for Research.
The ethical clearance for this protocol runs from 21-Nov-12 to 01-Feb-14.

Please forward your response to Manager, Research Ethics, Office for Research as per the details below.

Please refer to the attached sheet for the standard conditions of ethical clearance at Griffith University, as well as responses to questions commonly posed by researchers.

It would be appreciated if you could give your urgent attention to the issues raised by the Committee so that we can finalise the ethical clearance for your protocol promptly.

Regards

Manager, Research Ethics
Office for Research
Griffith University

Ethics Amendment

GRIFFITH UNIVERSITY HUMAN RESEARCH ETHICS COMMITTEE

29-Jan-2014

Dear [researcher]

I write further to your application for a variation to your approved protocol "NR: How the V Position is perceived, communicated and applied to the forehand groundstroke by coaches and players internationally." (GU Ref No: PES/45/12/HREC). This request has been considered by the Office for Research.

The OR resolved to approve the requested variation:
Variations requested as follows:

1) Change to distribution of information and consent materials (and questions) to allow for interview recruitment via direct contact over email.

1) Change to wording of interview questions.

NOTE: Copies of revised materials provided.

This decision is subject to ratification at the next meeting of the HREC. However, you are authorised to immediately commence the revised project on this basis. I will only contact you again about this matter if the HREC raises any additional questions or comments about this variation.

Regards

Policy Officer
Office for Research
Griffith University

Appendix C: Informed Consent Form and Information Sheet

<u>**Elite Coach Participant Consent Form:**</u>

Research Team:

Chief Investigator (1): ANONYMOUS School of Education and Professional Studies	Chief Investigator (2): ANONYMOUS Dean of Research (Health) Griffith Health Executive	Student Researcher: [researcher] School of Education & Professional Studies and School of Rehabilitation Sciences Doctor of Philosophy Candidate

By signing below, I confirm that I have read and understood the information package and in particular have noted that:

I understand that my involvement in this research will include answering some interview questions over email that will assist in analysing the coach-athlete relationship and athlete perception whilst performing the tennis groundstroke.

I understand that my involvement in this research also includes the research team collecting official publicly available data on previous coaching relationships that will help in the identification of effective coaching pedagogy and effective coach-athlete relationships.

I have had any questions about the Elite Coach Questionnaire and the collection of correspondence answered to my satisfaction.

I understand that there are no risks involved.

I understand that there will be no direct benefit to me from my participation in this research (I will be privy to the results nonetheless).
I understand that my participation in this research is voluntary.

I understand that if I have any additional questions I can contact the research team.

I understand that I am free to withdraw at any time without comment or penalty.

I understand that I can contact the Manager, Research Ethics, at Griffith University Human Research Ethics Committee on [phone] (or [email]) if I have any concerns about the ethical conduct of the project.

I agree to participate in the project.

Name:	
Signature:	
Date:	

Elite Coach Information Sheet:

Research Team:

Chief Investigator (1): ANONYMOUS School of Education and Professional Studies	Chief Investigator (2): ANONYMOUS Dean of Research (Health) Griffith Health Executive	Student Researcher: [researcher] School of Education & Professional Studies and School of Rehabilitation Sciences Doctor of Philosophy Candidate

Why is the research being conducted?

The current research project is being conducted ultimately to contribute to athlete performance and coaching optimisation in respect to the theorised V Position and its application. The research intends to identify the quintessential forehand groundstroke that enables athletes to achieve a fluid groundstroke whilst increasing their power production at impact with the ball. In order to reach the research objective, it is necessary to look at current elite performance and the coaching pedagogy that has enabled the elite to reach such heights. It is therefore the intention to use the knowledge of the elite, the knowledge of their respective coaches, to contribute towards the advancement of tennis and the increase in power-play during competition.

The necessity to analyse the elite is seen as crucial to the development of the potential advancement of the game, contributing towards the development of the younger generation and how their forehand groundstroke is developed. Additionally, it is of interest to the research to identify effective measures that will allow junior athletes to excel and increase their performance, increasing the bar for entering the competitive ranks on the WTA and ATP tours respectively.

The research intends to additionally impact current coaching pedagogy trends universally in respect to communicating movement and how athletes perceive the V position and apply the information they have interpreted to their performance during competition. The ability to effectively educate athletes and communicate the V position can be transferred across fields and contribute towards effective coaching practices collectively whilst also contribute to enhance athlete performance, specifically power out-put whilst maintaining accuracy, and doing so with ease – not compromising the ordinary movements of the body, rather using the athletes regular movements to enhance their performance.

What you will be asked to do?

As the coach, you will be asked to answer some interview questions over email.

The basis by which participants will be selected or screened:

It is within the scope of this research to access the highest level of the coaching profession and the elitist tennis players. For this to occur, the participant margin has been cut off at a WTA or ATP player ranking of 1-200. This range ensures that fifty-percent of potential participants (approximately 100 participants) are (or have been) Grand Slam contenders, therefore competing at the highest level in the sport of tennis.

The respective coaches are credited for player performance at this level, and it is the intention to ensure the players and their respective coaches are both participants to allow for perception and communication analysis between the coach-athlete relationship. Participants will be recruited within the specified range due to their unique set of skills (players) and their superior knowledge base (coaches).

The expected benefits of the research:

The research objective is to find effective means by which the coach-athlete relationship is at its strongest and to provide an outline for coach education into what creates an effective coaching pedagogy. Additionally, an effective pedagogy will align with athlete performance and their perception of this pedagogy and if it transfers into their performance and direct results. Ultimately, the aim of this research is to identify a specific movement pattern (the V Position) that is effectively communicated by the coach, performed by the player whilst increasing performance outcomes.

Risks to you:

There are no risks associated with your participation in this project.

Your confidentiality:

This project will collect data that is identifiable, for example, emails and individual questionnaires. This identifiable data will be converted to "de-identified" data, in that individual questionnaires and participants will be recorded with a coding system, not by name. There will be no identified individuals in the reporting of this project.

Individual questionnaires will be destroyed in accordance with Griffith University data storage policy once findings have been collated across multiple questionnaires. Only coded responses will remain.

Your participation is voluntary:

Your participation is completely voluntary. You may withdraw from the project at any time. If you do withdraw from the project, in no way will your participation in the research project be affected or compromised.

Questions / further information:

Project enquiries should be directed to Chief Investigator 1, Chief Investigator 2 or Griffith University Student Researcher [researcher].

The ethical conduct of this research:
Griffith University conducts research in accordance with the *National Statement on Ethical Conduct in Human Research*. If you have any concerns or complaints about the ethical conduct of the research project please contact the Manager, Research Ethics.
Feedback to you:

The collated findings of the questionnaires will be available to you as a participant. Please contact the project team to request a written copy of the findings.

Privacy statement:

The conduct of this research involves the collection, access and/or use of your identified personal information. The information collected is confidential and will not be disclosed to third parties without your consent, except to meet government, legal or other regulatory authority requirements. A de-identified copy of this data may be used for other research purposes. However, your anonymity will at all times be safeguarded. For further information consult Griffith University's Privacy Plan at http://www.griffith.edu.au/about-griffith/plans-publications/griffith-university-privacy-plan.

Consent for Questionnaires:

Returning the completed questionnaire is normally regarded as consent to participate in the questionnaire only. Your written consent is required for other data to be collected, such as correspondence between you and the research team.

On behalf of the Research Team, thank you for your participation in this project.

Appendix D: Email Correspondence

Email contact with the WTA tour
LinkedIn message

WTA CONTACT ANONYMOUS

31/10/2012

To: [researcher]

Hello Ashley,

Thanks for your enquiry regarding your research project.

As you have already approached the WTA, ATP and ITF about your proposal I expect the best way for you moving forward would be to consult with the Australian coaching community. I suggest that you contact Tennis Australia as they may be able to assist. Their coaching department I expect would be the most appropriate: "email"

They are also running a coaches' conference in January and this usually attracts large numbers of coaches, so that may be a suitable forum for you.

Good luck with your project.

Regards,

WTA
Contact Anonymous [Senior Director, Athlete Assistance]

Email contact with the ATP tour

On 10/16/12 4:17 PM, [researcher] wrote:

Dear ATP First Contact,

I hope this email finds you well. I would like to ask you how you would recommend accessing coaches on the ATP tour? I am conducting some research for my Doctor of Philosophy studies (at Griffith University, Australia), and would like coaches with players inside the top 200 to complete a questionnaire (online), and if possible, to have a brief chat with them (the coach) during one of the larger tournaments (where most would be at the same time) or via Skype. Also, I would like to know how I would be able to get the ATP tour to endorse my research - purely by allowing myself to observe play at the larger tournaments where these players would be competing and I would be more than willing to share my findings with the tour. Any information you may be able to provide would be greatly appreciated.

Many thanks,

[researcher]

Date: Monday, 22 Oct 2012 17:37:38 +0000

Subject: RE: **Research on behalf of PhD Candidate, Australia LinkedIn**

ATP First Contact Anonymous has sent you a message.

Subject: RE: Research on behalf of PhD Candidate, Australia

Ashley,

Thanks for your message. Your Doctor of Philosophy studies sounds very challenging. Coaching is a big part of a tennis players success and it would be interesting to know more. ATP CONTACT is the person within ATP who deals with coaches and I recommend you to send him an e-mail. He is also the tournament director of the ATP Tour World Finals in London.

Best regards

ATP First Contact Anonymous

Sent: Monday, October 22, 2012 10:38 PM

To: ATP CONTACT

Subject: RE: Research on behalf of PhD Candidate, Australia

Dear ATP Contact Anonymous,

I have recently been talking to "First ATP Contact" in relationship to accessing coaches on the ATP tour, and he gave me your name and said you may be willing able to assist.

My name is [researcher] and I am a PhD Candidate at Griffith University, Australia. I am looking for 100 coaches and their respective coaches on either the WTA or ATP tours with their current player ranked inside the top 200. As part of my research, I am after the coach to complete an online survey, and their respective player to complete a different online survey (the player currently being worked with). I am hopeful that these surveys would be completed by the end of January 2013 – they will take roughly 20 minutes to fill in, however once I send the link through, the coach will have roughly 2 months to complete the survey and can come back to it at any time.

In addition to the survey, I am hoping to interview the coach which will take roughly 5-10 minutes. I would like to do this face-to-face during a tournament, to be the most convenient, however if this is not possible, am happy to do it either via Skype or via an interview link (similar to the survey). It would be ideal if this could take place during the month of January 2013 at some stage.

This is all I am after from the coach and the player. From here, I will purely be observing the respective tour player during performance/competition so am hopeful that this will happen during the Australian Summer Series at some stage, alternatively, if this does not happen, tournaments up until and including through March 2013.

All participants will remain anonymous and all participants will be notified of the results. So what am I looking at? My research is looking at ways to enhance player performance – both biomechanically and communicatively at the elite level. Why do I want/need your help? The basis of my research and the results are based on coach opinion and player opinion – which will be only viewed by myself (and placed into my research where applicable); which it is then up to me to observe during competition if this happens i.e. if the coach tells me their player is working on 'x' then I may observe if they execute 'x' during a certain movement performed in their match.

The ATP would be of tremendous help if I was able to either A) get access to coach emails to ask them if they wouldn't mind participating (with their respective players), B) if the ATP was able to send out the link to all coaches (and their respective players) when appropriate - and potentially post an email through to them with the research being conducted and that their opinions are going to count (all participants will remain anonymous).

If you are able to let me know ATP Contact Anonymous if the ATP is able to help in any way, or be of assistance in allowing myself to gather this information, any help and assistance would be greatly appreciated. Again, for the players, it's purely an online survey they can do at their leisure (by February) and for the coaches, the same applies, with an additional interview during the month of January (perhaps after a practice or whenever is convenient). Ultimately, I am hopeful that as a collective, I can contribute to the game of tennis by using coaches on the ATP tours' opinion (therefore knowledge), in establishing an increase in player performance.

Thank you for your time ATP Contact Anonymous, if you could let me know if the ATP tour is able to help, this would be greatly appreciated.

Kind Regards

[researcher]

RE: Research on behalf of PhD Candidate, Australia

ATP CONTACT ANONYMOUS

26/10/2012

To: [researcher]

Ashley,

How are you? It is nice to meet you.

As you can imagine, there are quite a few requests to contact players and coaches in our tour. The reality is that many people do not really understand the dynamic of our organization. First of all, our players are not employees of ours, they are independent contractors that own 50% of our tour (the other 50% is owned by the tournaments). As such, we are not at liberty to give their personal information to someone outside the ATP. Furthermore, the coaches are an integral part of the tour but they are also independent contractors with no direct link to the ATP. They are hired by the players and while we deal with them daily, we are also not at liberty to give their personal information.

So, while I would like to help, I do not have the capability to this on my own.

I hope you understand

All the best,

ATP Contact Anonymous

Email Contact with Brisbane International

Sent: Thursday, 22 November 2012 10:19 AM

To: Brisbane International

Subject: Player Specific Research at the Brisbane International - PhD Candidate

Dear Brisbane International,

My name is [researcher] and I am a PhD Candidate at Griffith University, QLD - and am conducting tennis specific research in ways to optimise player performance. In order to collect the necessary information I need, I require coaches, and their respective players, to complete an online survey (I can make this link available to you). The target level that my research is focusing on is players (and their coaches) inside the top 200 on the WTA or ATP tours.

The Brisbane International is not only the Universities 'home' tournament, in respect to location, but the calibre of athletes competing in the Brisbane International is exactly what my research requires. I was going to, as normal, attend the Brisbane International as a spectator and be hopeful to briefly chat to some coaches and hopefully they wouldn't mind completing the survey - I would purely just give them a card with some information of where the survey is (the web address) and for the coach to discuss this with their athlete as I do not want to intrude or distract players by any means.

I would like to ask however, as it would make my research a lot more successful, if the Brisbane International would allow myself to post a flyer at the venue, or in the player/coaches lounge, or potentially when the players and their coaches check in, for the flyer/card to be handed to them if this is an option.

I currently have a few contacts who are coaches with players competing in the Brisbane International, however as I am after a large quantity (30+) of players to complete the survey, I would like to ask if the Brisbane International would be able to assist myself in some manner. The card/flyer does not need to necessarily go to the top 10 players, it is any player inside the top 200 - if all players, including qualifiers, were given a card/flyer, then I would be very hopeful that they would complete the survey. Potentially there is a coaches lounge where all coaches can receive the card/flyer as this would be preferable, as understandably, the players I am sure are barraged with a lot of marketing information and commitments.

I would appreciate any assistance the Brisbane International Team is able to offer.

Thank you kindly.

[researcher]

From: Brisbane International Contact

Date: Wednesday, 28 Nov 2012 09:23:41 +1100

Subject: FW: Player Specific Research at the Brisbane International - PhD Candidate

Ashley

Thanks for your email.

I would be happy to consider a request to conduct research. I would firstly need some further information re the study – I am happy to look at the online survey, however I feel that a quick summary re the questions and purpose would be of more benefit so we can forward to the WTA and ATP Tour Supervisors for the Brisbane International. I would be guided by them to the appropriateness of this research being done on site.

I will await further information from yourself to send to our Tour Supervisors and Tournament Director.

Thanks and regards

Brisbane International Contact (Projects Manager)

Sent: Wednesday, 28 November 2012 9:03 AM

To: Brisbane International Contact

Subject: RE: Player Specific Research at the Brisbane International - PhD Candidate

Dear Brisbane International Contact

Thank you kindly for your email.

To give you more information on the study:

The study's premise is based on 1) enhancing the coach-athlete relationship on an international scale, allowing governing bodies to implement the found aspects into their already running programs and allow the WTA and ATP access to new sports science specific research that they can make available to the global coaching fraternity. The participants sort are those inside the top 200 on the WTA and ATP tours with their respective coaches. It is the intention of the study to view the 'elite' and for the elite coaches and their respective players to complete a survey (online at their convenience), allowing for the study to work backwards so to speak. The study here is twofold - backwards in respect to identifying and implementing strengths of the elite into national coaching programs, and enhance - by utilising the gathered information to enhance the elite players' already elite standard - this links into the second aspect; 2) optimising player performance by increasing power out-put through the forehand groundstroke; this analysis will contribute towards a more efficient playing pattern that the already elite will be able to use, with the theorised movement (of the study), and that will additionally roll back into the coach-athlete relationship and is hopeful to be a part of the coaching foundations worldwide.

The big element however of interest is 1) the coach response, and 2) the athlete response of the survey (there is a specific survey for the coach and another for the athlete). It has been observed that the elite are elite for a reason and that their opinions are of extreme value, and it is this that the survey is after - pure coach and player opinion of the coach-athlete relationship, and their performance. The elite coaches and players therefore will be a part of an international research project in hope of being able to contribute to international coaching bodies, not just Australian - their own and other countries. All coaches and players will not be identified and will remain anonymous - they will have the option to leave their name for follow up questions, however at all times their privacy will be of the highest importance and names only available to the research team for contact purposes. Once contact has been finalised, all names will be de-identified (therefore unknown).

It would be a great opportunity if the coaches and their respective players were asked to complete the survey in their own time at the Brisbane International for the benefit of the research - I'm hopeful at some time by the start of February 2013 - whether it be on their plane ride or whilst at the venue on a quiet night etc. I will make business cards available with the survey link on the card to ensure the link can be placed in their pocket for safe keeping. If it is alright, it would be very appreciative it either the players and coaches could receive a business card and be briefly informed that it is purely about research and their own opinions or if the Tour Supervisors respectively prefer to hand the card out, that would be equally helpful. The survey will inform them of the purpose when they click on the link - it has been made for convenience and to inform the coaches and players opposed to having to be stopped to talk to. If the first option of delivery isn't ideal, then I would equally appreciate (or in addition to), talking to coaches at the event and handing out the card with the survey link (however I would be trying my luck at who to talk to as personally will not know too many of the coaches) - I do feel as though the former option would be more kindly viewed as I do not wish to appear intrusive.

I hope this information is sufficient Brisbane International Contact and I am more than happy to answer any additional questions in hope that the Brisbane International entrants complete a survey that will ultimately contribute to their sport and the way it is coached, and a way to optimise performance.

Thank you for your time Brisbane International Contact.

Kind Regards

[researcher]

From: Brisbane International Contact

To: [researcher]
Date: Wednesday, 5 Dec 2012 17:21:45 +1100

Subject: RE: Player Specific Research at the Brisbane International - PhD Candidate

Hi Ashley

I have been in touch with both the WTA and ATP Tours regarding your request. Unfortunately both the Tours as well as our Tournament Director feel that this type of research is not conducive to our event.

I apologise we couldn't be of any further assistance.

Thanks and regards

Brisbane International Contact

Sent: Wednesday, 5 December 2012 8:54 PM

To: Brisbane International Contact

Subject: RE: Player Specific Research at the Brisbane International - PhD Candidate

Dear Brisbane International Contact

Thank you very much for your email. May I please ask if there is anyway, or possibility, that I may then with the given decision, be able to place a Business Card in any shape or form that the players would then have access to the survey? Alternatively, is there a way that the respective coaches of the athletes would be able to receive the Business Card (inclusive of the survey link)? I would really be very thankful and appreciative of any help I may be able to get in order to get all the Brisbane International players and their respective coaches access to the survey link - even if this means only those in the qualifying rounds and/or the early main draw if this appears to be problematic.

Also, if there is any way I may be able to persuade or shape my research that meets the Tour Supervisors' and Tournament Directors best interests, I would look at building a specific report on the participants etc (or similar) that would purely be available to those mentioned.

Thank you Brisbane International Contact, I hope to hear potentially good news from you - enjoy your week.

Kind Regards

[researcher]

RE: Player Specific Research at the Brisbane International - PhD Candidate

Brisbane International Contact

12/12/2012

To: [researcher]

HI Ashley

As I have already made the request to both the WTA and ATP Tours, I respect that their decision is final.

I apologise I could not be of any further assistance.

Thanks and regards

Brisbane International Contact

Email contact with Tennis Australia

Sent: Saturday, 6 October 2012 10:23 AM

To: Tennis QLD Initial Contact

Date: 10/06/2012

Subject: Research on behalf of PhD Candidate, Australia

Dear Esteemed Colleagues, Professional Coaches and Players, and Tennis Experts,

My name is [researcher] and as you would all know from my profile I am studying my PhD (Doctor of Philosophy) in relationship to tennis and enriching the game. In order to conduct my research, I am asking for your help and assistance.

I am after 100 x Professional Tennis Coaches and their respective players with rankings inside the top 200 on either the WTA and/or ATP Tours. I am hopeful that during the Australian Summer Series I will be able to ask the coaches (potentially yourself) several questions (your opinions), and players to complete a brief survey and to have a brief chat (no more than 5 minutes) - preferably after a practice session.

To all those who are responsible for the organisation of these events and their operations, I would like to ask permission also to conduct my research - gaining the support from national bodies, whether that be Australian and/or others, would be very much appreciated.

The research is centred on a global perspective, therefore requires the international coaching fraternity and their players (all whom will remain anonymous) that will contribute towards a better way of developing the junior and elite tennis player and coaching practices.

I understand I have sent this email to several persons in multiple fields. If you are able to help, if you do not mind answering a few questions, if you can participate, if you would like to endorse this research (on behalf of Griffith University, QLD, Australia), or if you have any questions, please email me - your help and participation, and endorsement will be greatly appreciated.

This is a lot to ask from so many persons, however as you would all understand, tennis is a global sport with different hierarchy's but we are all working under the one arena - Tennis; and I hope with this research it can contribute to enriching the game on many levels.

Thank you all for your time, and if you require further information from Griffith University or myself, I am more than happy to provide it and/or direct you to my university head of school. I would appreciate it if you could respond so I can let you know dates (through December to January - and if you are a coach and/or player, I will come to you - if you are a coaching body, I will also come to you if needed for your endorsement).
Kind Regards from QLD, Australia,

[researcher]

From: Tennis QLD Initial Contact

Date: Sunday, 7 Oct 2012 12:15:01 +1100

Subject: RE: Research on behalf of PhD Candidate, Australia

Hi Ashley

All requests such as these need to be directed to TA Authority Contact who is the High Performance Manager at Tennis Australia.

As the leader of the High Performance, Sports Science and Medicine areas for Tennis in Australia, TA Authority Contact has processes in place that cater for your request.

I have copied him on this email.

Tennis QLD initial contact

Sent: Tuesday, October 09, 2012 10:37 AM

To: TA Authority Contact

Subject: Research on behalf of PhD Candidate, Australia

Dear TA Authority Contact,

My name is [researcher] and I am studying my PhD at Griffith University (QLD) looking at both the biomechanics and communication of the tennis groundstroke. I am in the process of reading a lot of your work and know that your research and assistance will/would be very helpful. I emailed Tennis QLD Initial Contact who said you are the best person to speak with in this respect - I am looking at getting the support of Tennis Australia to conduct my research during the Australian Summer Series. I am after 100 x Professional Tennis Coaches and their respective players with rankings inside the top 200 on either the WTA and/or ATP Tours and am hopeful that I will be able to ask the coaches several questions, and players to complete a brief survey and to have a short chat (no more than 5 minutes) - preferably after a practice session. I am not 100% sure how to proceed with getting Tennis Australia on board and to be granted permission to conduct my research, and to get coaches and players to answer a few questions. May I please ask for your guidance in this respect and how I should proceed. If you need confirmation about my candidature, [former supervisor] is one of my supervisors, which I am sure you know and he will be more than happy to confirm with yourself my research purpose.

Thank you for your time TA Authority Contact.

Kind Regards

[researcher]

From: TA Authority Contact

Date: Thursday, 11 Oct 2012 19:35:08 +1100

Subject: RE: Research on behalf of PhD Candidate, Australia

Hi Ashley

Thanks for the email.

We typically deal with research requests that involve our Australian players and make decisions to support those requests (or otherwise) based on the return or potential return of the proposed research against our strategy priorities but also in light of the number of concurrent projects that our players/coaches may be involved with.

However, the nature of your request relates to a playing group with which we have far less involvement. That doesn't make it any less meritorious but just far more complicated.

Consequently, your research will be subject to the same criteria as described above but also vetted further by the stakeholders most involved in hosting the tournaments.

Notwithstanding that the proposed outcomes would need to be meaningful to tennis in this country, the Australian events take the playing / tournament experience very seriously so I suspect that there would be some healthy resistance to undertaking any (and not just your) research initiatives at a tournament.

Nevertheless, if you are able to provide more detail about what you had planned, I will raise it with those concerned. I will not get you to complete one of our research applications at this point as I would hate for you to put in all that work if the response is unlikely to be favorable.

Any questions, don't hesitate to call on [phone number]

Cheers
TA Authority Contact

Sent: Tuesday, October 16, 2012 12:21 PM

To: TA Authority Contact

Subject: RE: Research on behalf of PhD Candidate, Australia

Dear TA Authority Contact

Thank you for your reply, it's very much appreciated. I understand the limitations that it appears I am up against however I can attempt to mold my research to be of greater benefit to tennis in Australia contrast to other nations.

Ultimately I am after a large number of players at an elite level and the Australian Open would be a great stage to collect the data I need. If I was able to have access to a number of Australian professional players, then I would be happy to constitute my research on these players and their respective coaches alone. I am however after 50+ players and their respective coaches, and I would like more so around the 80+ mark (up to 100) - including both males and females inside the top 200 on either the WTA or ATP tours respectively.

To the best of my knowledge I am unsure if Australia presently has this number of players inside the top 200, however combining both tours, potentially out of 400 players we may have 20%. If there is more, or if I increase the margin to top 300 and this is inclusive of more Australian players, then I would discuss this with my supervisors and would be prepared to widen the gap. If not, perhaps you would be able to recommend a way of myself accessing the information I require.

In order to collect my data TA Authority Contact I understand that during tournament play it is on my behalf a good opportunity, however for the players - not so much. This said, I do not necessarily want to talk with the players as I understand this may be quite imposing, however would like a brief chat with their coaches (a brief chat with the player would be a bonus).

I do wish to observe the coaching session; however I have put this in place of talking with the player to avoid any interruptions to their training. Therefore ultimately I am looking for 100 players to complete the questionnaire I am fine-tuning and for their coaches to also complete the questionnaire within 1-8 weeks in attempt not to impose too much. Potentially it is best to contact the WTA and ATP tours, however I would be very appreciative to have Tennis Australia's approval.

I do however wish to observe the players' respective matches, and therefore would be helpful if I had access to these matches. I have attempted to ensure my research design is the least imposing on the player, and will be happy outside tournament time to conduct this research as well (the player and coach do not have to complete the questionnaire at the tournament) - if given permission to do so; again my objective is not to come across as being intrusive, purely to ask some questions, get the coach and athlete to complete a questionnaire, and to be able to observe the coaching session (only so I can hear the conversation from a far), and then to observe when the player is competing. If it would be uncomfortable for myself to be present, perhaps I could place a small taping device so I could play back the coach-athlete communication and set up a time for a brief chat with the coach (outside the venue if it is more suitable).

TA Authority Contact, any guidance and assistance you can provide will be greatly appreciated.

Thank you for your time.

Kind Regards

[researcher]

From: TA Authority Contact

Date: Saturday, 20 Oct 2012 23:05:55 +1100

Subject: RE: Research on behalf of PhD Candidate, Australia

Hi Ashley,

Unfortunately, we won't be able to facilitate what you have described below.

I have forwarded your email on to a few key internal stakeholders. As expected, they are particularly selective with the type of work that they support with the cohort that you have in mind and, at this stage, your proposal does not quite fit the bill.

From an academic perspective, I would encourage you to continue to work with [anonymous] in refining your scope.

I am on leave next week but feel free to call afterward to discuss this or the below in more detail.

Cheers

TA Authority Contact

Email contact inquiring about Media Pass for the Australian Open

Sent: Monday, October 22, 2012 9:51 AM

To: TA Authority Contact

Subject: RE: Research on behalf of PhD Candidate, Australia

Hi TA Authority Contact,

Thank you for your reply. From your first email, I gathered that this may be the case. I will endeavor to contact and work with coaches on the WTA and ATP tours, it is however unfortunate that Tennis Australia isn't able to assist. Nevertheless, I do appreciate the information you have been able to provide. On another TA AUTHORITY CONTACT, would you be able to advise myself if it would be at all possible to get a media pass to the Australian Open and/or other events during the Australian Summer Series? I'm looking at getting access to court-level vision in order to observe with more clarity and also take potentially photographs that will contribute to the research. If you could let me know at all if this is possible, that would be greatly appreciated.

Thank you again TA Authority Contact, I appreciate your help.

Kind Regards
[researcher]

RE: Research on behalf of PhD Candidate, Australia

TA Authority Contact

4/11/2012

Hi Ashley,

To be honest, the media pass will be very difficult to come by as they are for dedicated/certified stakeholders from the media.

Again feel free to give me a call.

Cheers

TA Authority Contact

Email contact with Tennis Australia coach

On 03/06/13 3:55 PM, [researcher] wrote:

Dear Anonymous Coach

I hope you've been well and thank you for the connection. You may (or may not) remember meeting a few years back when I was the Director of Tennis at Former Coaching Centre and also when I was coaching at the Other Former Coaching Centre.

As you may have read from my profile, I'm currently doing my PhD (research) on the forehand groundstroke in elite tennis players; I'm also looking at the opinions of elite tennis coaches. For this, I've designed a survey that takes 15-20 minutes to complete.

I attended the Brisbane International and Australian Open earlier in the year, but we didn't get the chance to meet - I tried my best to introduce myself to most coaches, however as I'm sure you would know, many of you are extremely busy at this time.

If it's not too much trouble, may I ask if you wouldn't mind completing the survey? I'm trying to get as many coaches and players to complete the survey as possible who are of the 'elite' class. Players inside the top 200 on either the ATP or WTA tours, and coaches who are working with these players. I'd be very very grateful if you could take 15-20 minutes out of your busy schedule Scott in the next week to complete the survey, or if you're not working with any top players at present, perhaps if it's not too much trouble I could ask if you wouldn't mind sharing this survey within your tennis network and hopefully some of your friends/former players/fellow coaches wouldn't mind completing the survey? The link to the coach survey is: [confidential]

And any players you may be able to get to complete the survey (there is a separate one for players and coaches) would really help my research (looking at optimising the forehand groundstroke and allowing the elite to develop more power). The link to the player survey is: [confidential]
I should note that the survey is anonymous, so I won't know who's who - there is the option at the end however to leave your details if you don't mind any potential follow-up questions at a later date.

Thank you Anonymous Coach, and thank you for joining my network - if you're ever on the Gold Coast, Australia, and would be willing to sit down for a coffee, I'd be much obliged.

Kind Regards

[researcher]

Date: Friday, 8 Mar 2013 03:31:18 +0000

Anonymous Coach has sent you a message.
Date: 3/08/2013

Subject: RE: Tennis Survey for PhD - ATP and WTA tour coaches

Hi Ashley,

Hope you're doing well! I'm not sure if my last response got sent so I'll reply again... I'm happy to do the survey, no problem. Is there a combined player/coach survey, or which one do you want me to do as I could fill on either?

Cheers

Anonymous Coach

Date: Friday, 8 March 2013 4:52 PM

To: Anonymous Coach

Subject: RE: Tennis Survey for PhD - ATP and WTA tour coaches

Dear Anonymous Coach

That would be fantastic and I'd truly appreciate it! And my apologies for not receiving your last email.

If you're currently working with a top 200 player (WTA or ATP) the coach survey would be great: [confidential]

And if it's not too much trouble for your player, if they could fill in the athlete survey it would be very helpful: [confidential]

Thank you again Anonymous Coach, and if you have any fellow coaches who don't mind taking the time to complete the survey, I'd really appreciate it. All coaches and players who provide their details will get the results from the survey (collective and what they mean) when they've been finalised (if interested). I'm hoping to get 50 x coaches and players so any help you'd be able to provide would be really truly appreciated.

Thank you Kindly Anonymous Coach,

[researcher]
From: Anonymous Coach

Date: Friday, 5 Apr 2013 13:51:36 +1100

Subject: Re: Tennis Survey for PhD - ATP and WTA tour coaches

Hi Ashley,

As I understand it, "TA Authority" have requested that you follow different channels so that all of us can benefit from your work.

Feel free to follow up with "TA Authority" directly.

Apologies, but at this time I can't complete the survey you have requested.

All the best!
Regards,

Anonymous Coach

Date: Friday, 5 April 2013 1:13 PM

To: Anonymous Coach

Subject: RE: Tennis Survey for PhD - ATP and WTA tour coaches

Hi Anonymous Coach

Thank you for your email. "TA Authority" wishes for my research to purely focus on Australian coaches and players, however this isn't possible to get an international perspective of the top coaches and players due to the variety of nationalities involved in elite tennis on both the WTA an ATP tours. I'm not sure how this stops you from completing the survey as it's in no way associated with Tennis Australia - and I have never exchanged emails with "Australian Open Tournament Director" (as much as I would like to).

All persons who participate in the study will receive the results - including any coaches directly associated with TA. I have a handful of Australian coaches who have completed the survey as all participants remain anonymous and in all honestly, TA is unable to know who has due to Griffith University's Ethics Protocol.

As a former elite player Anonymous Coach, you would know that players are from diverse nationalities, as are their coaches - and in order to get an overall understanding I need to incorporate all nationalities; although this doesn't guarantee others will participate, I must try for the value of my research. I actually will be presenting at the "Conference", recently being accepted, on behalf of Griffith University - specifically addressing the coaching element my research addresses. I'm telling you this in hope that you can see how I'm trying to reach an international audience. Tennis Australia has offered myself no funding nor collaboration, so I am in no way affiliated with TA nor is my research - it is purely individual and separate.

I'm hopeful you will change your mind in completing the survey; again Anonymous Coach, all persons remain anonymous and are unable to be identified in the research at no stage.

Thank you again for your email Anonymous Coach.
Kind Regards

[researcher]

From: Anonymous Coach

Date: Tuesday, 7 May 2013 11:49:20 +1000

Subject: Re: Tennis Survey for PhD - ATP and WTA tour coaches

Hi Ashley,

Sorry for the slow reply…. I have copied "TA Authority" in on this reply.

As I understand it, "TA Authority" are less concerned with a unilateral focus on Australian players and coaches but rather that all research requests (of which we receive many) are coordinated and that we try to leverage projects to extract as much value as possible out of them. It is due process on our part and in my experience, this is often best achieved through collaboration.

I would encourage you to touch base with "TA Authority" to ascertain how the interview might be shaped to gain more interest from our playing group as compared to just myself. Great news regarding your acceptance to present at Conference.

Cheers

Anonymous Coach

Date: Tuesday, 7 May 2013 11:54 AM

To: Anonymous Coach

Subject: RE: Tennis Survey for PhD - ATP and WTA tour coaches

Dear Anonymous Coach,

Good to hear from you and thank you kindly for your reply. From what's been said in the past, I would have to alter my project in order to get TA behind my research, however that only opens the gateway to a select number of coaches, like yourself; on the other hand, without altering my project, I allow it open for other national bodies - which is what I need to do to keep it on an international scale. As much as I would of loved to have TA's support, I'm not prepared to neglect other nations due to the multi-national presence of tennis globally. It's unfortunate that you're unable to be involved, as that would have been wonderful - but I'm sure you understand Anonymous Coach why I wish to keep the project on an international scale.

Thank you again kindly for your reply Anonymous Coach and kind words - all the best.

Kind Regards

[researcher]

07 May, 2013

FROM: ANONYMOUS COACH

Okay Ashley – no worries, I understand and apologies I couldn't be of assistance this time around.

All the best!

Regards

Anonymous Coach

Appendix E: Work Log

Note: Names of tennis players and their associates have all been replaced with *anonymous tennis contact*.

DATE & TIME:	ACTIVITY:	LOCATION/ EVENT:	SUBJECT OBSERVED:	DATA COLLECTED: Y/N
DAY 1/10 28.12.12 10am - 4pm at Venue (8am - 6pm Travel)	Survey Distribution	Brisbane International, QLD	NA (Qualifying Round)	6 x Survey's handed out. Notable hand out to *anonymous tennis contact** Contact details received from Coach based in Florida.
DAY 2/10 29.12.12 9.50am - 3.40pm at Venue (8am - 5.30pm Travel)	Survey Distribution	Brisbane International, QLD	3 x Matches (6 x Players)	4 x Surveys handed out. Notable handout to 3 *anonymous tennis contacts*
DAY 3/10 30.12.12 9.40 - 4.40pm at Venue (8am - 6.30pm Travel)	Survey Distribution + Observational Analysis	Bris. Int. QLD	4 x Matches (7 x Players)	3 x Surveys handed out. Notable handout *anonymous tennis contact*
DAY 4/10 31.12.12 9.45 - 9pm at Venue (8am - 11pm Travel)	Survey Distribution + Observational Analysis	Bris. Int. QLD	5 x Matches (9 x Players)	2 x Surveys handed out (plus additional to *anonymous tennis contact* to hand out to other coaches). Notable handout to 2 *anonymous tennis contact's* coaches

DATE & TIME:	ACTIVITY:	LOCATION/ EVENT:	SUBJECT OBSERVED:	DATA COLLECTED: Y/N
DAY 5/10 01.01.13 10.15 - 2.15pm at Venue (9am - 4pm Travel)	Observational Analysis	Bris. Int. QLD	2 x Matches (4 x Players)	NA Note: had to leave earlier than intended due to lower limb injury.
DAY 6/10 02.01.13 10.15 - 6pm at Venue (8.45am - 8pm Travel)	Survey Distribution + Observational Analysis	Bris. Int. QLD	4 x Matches (7 x Players)	4 x Surveys handed out (3 x Players, 1 x Coach) Handed out a handful to 2 x Security guards who said they may be able to pass them onto the players and coaches when they come off the practice courts.
DAY 7/10 03.01.13 11.15am - 4.15pm at Venue (9.15am - 6.15pm Travel)	Survey Distribution + Observational Analysis	Bris. Int. QLD	3 x Matches (6 x Players)	4 x Surveys handed out (3 x Coaches + 1 x Player) Note: coach and player approached together (*anonymous tennis contacts*) Notable handouts – 3 *anonymous tennis contacts* coaches
DAY 8/10 04.01.13 12.15 - 3.15pm at Venue (11am - 5pm Travel) - drove to venue.	Survey Distribution + Observational Analysis	Bris. Int. QLD	1 x Match (2 x Players)	7 x Surveys handed out (2 x Juniors, 1 x Coach with 2 x Players, 1 x Coach with 1 x Player) Notable handouts - *anonymous tennis contact's* Coach

DATE & TIME:	ACTIVITY:	LOCATION/ EVENT:	SUBJECT OBSERVED:	DATA COLLECTED: Y/N
DAY 9/10	no play (finals)	NA	NA	NA
DAY 10/10	no play (finals)	NA	NA	NA
Day 1/14 09.01.13 9.15 - 5.30pm at Venue (8am - 7pm Travel)	Survey Distribution + Observational Analysis	Melb.	6 x Matches (12 x Players)	6 x Coach 4 x Player
Day 2/14 10.01.13 9.45 - 5.45pm at Venue (8.10 - 7.20pm Travel)	Survey Distribution + Observational Analysis	Melb.	6 x Matches (12 x Players)	6 x Coach 1 x Player (+ 1 x wife of coach/physical trainer) Note: handed out a handful of survey's to *anonymous tennis contacts* - would like to know match stats for *anonymous tennis contact*.
Day 3/14 11.01.13 9.50 - 5.30pm at Venue (8.10 - 7.20pm Travel)	Survey Distribution + Observational Analysis	Melb.	6 x Matches (+ 1 x 1/2 Match) - (12 x Players + 1/2 of 2 x Players)	2 x Coach 1 x Player Notable handout to *anonymous tennis contact*
Day 4/14 12.01.13 10.10 - 4.30pm at Venue (9.10 - 6pm Travel)	Survey Distribution + Observational Analysis	Melb.	2 x Matches (2 x Players) (Play didn't start until 12.30pm - normally 10am)	3 x Coach 3 x Players Notable handout to 2 *anonymous tennis contacts*

DATE & TIME:	ACTIVITY:	LOCATION/ EVENT:	SUBJECT OBSERVED:	DATA COLLECTED: Y/N
Day 5/14 14.01.13 9.50 - 4.30pm at Venue (8.15 - 6.30pm Travel) Weather x Long lines to get in x	Survey Distribution. + Data Analysis	Melb.	4 x Matches (8 x Players)	1 x hitting partner (Aussie hitting partner - gave 4 x cards for coaches and player of *anonymous tennis contact*)
Day 6/14 15.01.13 9 - 5.40pm at Venue (7.30 - 7.10pm Travel) Gates don't open until 10am - to avoid long lines came early.	Survey Distribution + Data Analysis	Melb.	4.5 x Matches (9 x Players)	2 x Player 3 x hitting partners (3 cards handed out, then 4, then 4) 1 x coach (former player) - *anonymous tennis contact*. Notable handout to *anonymous tennis contact* (x 2) + hitting partner on *anonymous tennis contact's* court.
Day 7/14 16.01.13 9 - 5.15pm at Venue (7.20 - 7.20pm Travel)	Survey Distribution + Data Analysis	Melb.	5 x Matches (+ 1/2 of 1 x Match - 1 x Set)	7 x Coaches (inc. maybe 1 x hitting partner) 1 x Player 2 x hitting partner (each given x 4 cards) Notable handout to *anonymous tennis contact's* Coach and *anonymous tennis contact*

DATE & TIME:	ACTIVITY:	LOCATION/ EVENT:	SUBJECT OBSERVED:	DATA COLLECTED: Y/N
Day 8/14 17.01.13 9.10 - 3.20pm at Venue (7.35 - 5.20pm Travel)	Survey Distribution + Data Analysis	Melb.	3 x Matches (6 x Players)	6 x Coaches 2 x Players (1 x Aussie, 1 x American) Notable handout to 2 *anonymous tennis contact's* Coaches
Day 9/14 18.01.13 9 - 1pm at Venue (7.30 - Travel)	Survey Distribution + Data Analysis	Melb.	1 x Match (2 x Players) - left early (tired and foot pain)	7 x Coaches 1 x Trainer (w/ 4 x players on court) Notable handout *anonymous tennis contact's* Coach (same)
Day 10/14 20.01.13 (Saturday Off - no big matches) 9.20 - 2.10pm at Venue (7.45 - 3.45pm Travel)	Survey Distribution (no singles match)	Melb.	NA	4 x Coaches Notable handout to *anonymous tennis contact* (x2) *anonymous tennis contact* (x2), *anonymous tennis contact* (x3), *anonymous tennis contact* (x5) Coaches
Day 11/14 21.01.13 1 - at Venue (11.30 - Travel) No matches on outside courts to record and limited access to practice courts.	Survey Distribution (no singles match)	Melb.	NA	3 x Coaches (*anonymous tennis contact's* Coach + 1 x Men's doubles' Coach + 1 x women's doubles' Coach - they aren't ranked inside top 200 for singles, unsure about men, think they may have been, but both are for doubles) 1 x hitting partner / friend (*anonymous tennis contact* and Coach on court)

Bibliography

Alvesson, M., & Skoldberg, K. (2009). *Reflexive methodology: new vistas for qualitative research* (2nd ed.) London, UK: Sage Publications.

Ángel, G. M., Evangelos, T., & Alberto, L. (2006). Defensive systems in basketball ball possessions. *International Journal of Performance Analysis in Sport, 6*(1), 98–107.

Araya, J. A., & Larkin, P. (2013). Key performance variables between the top 10 and bottom 10 teams in the English premier league 2012/13 season. *University of Sydney Papers in HMHCE, 2*, 17–29. Sydney, Australia: University of Sydney.

Berge, A. M. (2014). How is the V Position applied and communicated in the forehand groundstroke by elite players and coaches internationally? Gold Coast, Australia: AM8 International.

Berge, A. M. (2017). The Science of Elite Performance: The World Awaits – *an infusion of science, education and communication: a complete learning structure on how elite coaches and athletes become the best in the world.* Gold Coast, Australia: AM8 International.

Biernacki, P., & Waldorf, D. (1981). Snowball sampling: problems and techniques of chain referral sampling. *Sociological Methods and Research, 10*(2), 141-163.

Callaway, A. J., & Broomfield, S. A. (2012). Inter-rater reliability and criterion validity of scatter diagrams as an input method for marksmanship analysis: computerised notational analysis for archery. *International Journal of Performance Analysis in Sport, 12*(1), 291-310.

Ciuffarella, A., Russo, L., Masedu, F., Valenti, M., Izzo, R. E., & De Angelis, M. (2013). Notational analysis of the volleyball serve. *Timisoara Physical Education and Rehabilitation Journal, 6*(11), 29–35.

Clemente, F., Couceiro, M., Martins, F., & Mendes, R. (2012). Team's performance on FIFA u17 world cup 2011: study based on notational analysis. *Journal of Physical Education and Sport*, *12*(1), 13–17.

Cohen, L., Manion, L., & Morrison, K. (2007). *Research Methods in Education* (6th ed.). Florence, KY: Routledge.

Cushion, C., Harvey, S., Muir, B., & Nelson, L. (2012). Developing the coach analysis and intervention system (cais): establishing validity and reliability of a computerised systematic observation instrument. *Journal of Sports Science, 30*(2), 203-218.

Davies, D., & Dodd, J. (2002). Qualitative research and the question of rigor. *Qualitative Health Research, 12*(2), 279-289.

Denzin, N.K., & Lincoln, Y.S. (Eds.)(2011). The SAGE handbook of qualitative research. Thousand Oaks, CA: Sage Publications.

Dogramaci, S. N., Watsford, M. L., & Murphy, A. R. (2011). The reliability and validity of subjective notational analysis in comparison to global positioning system tracking to assess athlete movement patterns. *Journal of Strength & Conditioning Research, 25*(3), 852-859.

Edwards, A. (2009). Qualitative research in sport management. Burlington, MA, USA: Routledge. Retrieved from http://www.ebrary.com

Elliott, B. (1991). Stroke production in tennis. *National Sports Research Centre* (Vol. 27, pp. 1–34). Bruce, Australia: Australian Sports Commission.

Elliott, B. (2002). Biomechanics of tennis. In P. A. F. H. Renstrom (Ed.), *Tennis* (pp. 1–28). Malden, MA: Blackwell Science.

Elliott, B., Fleisig, G., Nicholls, R., & Escamilia, R. (2003). Technique effects on upper limb loading in the tennis serve. *Journal of Science and Medicine in Sport, 6*(1), 76–87.

Gillet, E., Leroy, D., Thouvarecq, R., & Stein, J. (2009). A notational analysis of elite tennis serve and serve-return strategies on slow surface. *Journal of Strength &Conditioning Research/National Strength & Conditioning Association, 23*(2), 532–539.

Hennink, M., Hutter, I., & Bailey, A. (2011). *Qualitative Research Methods* (pp. 14–15). London: Sage Publications.

Hertz, R., & Imber, J. (1995). *Studying Elites Using Qualitative Methods*. Thousand Oaks, CA: Sage Publications.

Hughes, M. (2002). The application of notational analysis to racket sports. In A. Lees, I. Maynard, M. Hughes & T. Reilly (Eds.), *Science and Racket Sports II* (pp. 211–220). New York, NY: Routledge.

Hughes, M., & Bartlett, R. M. (2008). What is performance analysis. *The Essentials of Performance Analysis: An Introduction* (pp. 15). Oxon, UK: Routledge.

Hughes, M., & Franks, I. (2004). *Notational Analysis of Sport: Systems for Better Coaching and Performance in Sport* (2nd ed). New York, NY: Routledge.

Hughes, M., Hughes, M. T., & Behan, H. (2007). The evolution of computerised notational analysis through the example of racket sports. *International Journal of Sports Science and Engineering, 1*(1), 3–28.

ITF. (2011). *ITF Rules of Tennis*. London, UK: ITF Ltd.

James, N., Mellalieu, S., & Jones, N. (2005). The development of position-specific performance indicators in professional rugby union. *Journal of Sports Science, 23*(1), 63–72.

Jupp, V. (2006). Documents and critical research. In R. Sapsford & V. Jupp (Eds.), *Data Collection and Analysis*. (pp. 272–290). London, UK: Sage Publications.

King, N., & Horrocks, C. (2010). *Designing an Interview Study*. London, UK: Sage Publications.

Knudson, D., & Morrison, C. S. (2002). *Qualitative Analysis of Human Movement* (2nd ed.). Champaign, IL: Human Kinetics.

Kuntze, G., Mansfield, N., & Sellers, W. (2010). A biomechanical analysis of common lunge tasks in badminton. *Journal of Sports Science*, 28 (2), 183-191. DOI: 10.1080/02640410903428533

Kwan, M., & Rasmussen, J. (2010). The importance of being elastic: deflection of a badminton racket during a stroke. *Journal of Sports Science, 28*(5), 505–511. DOI: 10.1080/02640410903567785

Kwitniewska, A., Dornowski, M., & Hökelmann, A. (2009). Quantitative and qualitative analysis of international standing in group competition in the sport of rhythmic gymnastics. *Baltic Journal of Health and Physical Activity, 1*(2), 118–125.

Long, J. (2007). *Researching Leisure, Sport and Tourism – The Essential Guide*. London, UK: Sage Publications.

Lund, A. &, Lund, M (2013) Descriptive and inferential statistics. *LAERD Statistics.* Retrieved from https://statistics.laerd.com/statistical-guides/descriptive-inferential-statistics.php.

Lupo, C., Tessitore, A., Minganti, C., King, B., Cortis, C., & Capranica, L. (2011). Notational analysis of American women's collegiate water polo matches. *Journal of Strength & Conditioning Research/National Strength & Conditioning Association, 25*(3), 753–757. DOI: 10.1519/JSC.0b013e3181cc245c

Magill, R. A. (2007). *Motor Learning and Control: Concepts and Applications* (8th ed. ed.). New York, NY: McGraw-Hill.

McGarry, T. & Franks, I.M. (1996). In search of invariant athletic behaviour in sport: an example from championship squash match-play. *Journal of Sports Sciences*, 14:5, 445-456. DOI: 10.1080/02640419608727730

Mooi, E., & Sarstedt, M. (2011). *A Concise Guide to Market Research: The Process, Data, and Methods using IBM SPSS Statistics*. New York, NY: Springer.

Moyser, G., & Wagstaffe, M. (1987). *Research Methods for Elite Studies*. London, UK: Allen & Unwin.

Munroe-Chandler, K.J. (2005). A discussion on qualitative research in physical activity. *Athletic Insight: The Online Journal of Sports Psychology 7*(1), 67-81.

Ortlipp, M. (2008). Keeping and using reflective journals in the qualitative research process. *The Qualitative Report, 13*(4), 695-705.

Oxford Dictionary (2015). Optimal. Retrieved July 28, 2015, from http://www.oxforddictionaries.com/definition/english/optimal

Packer, M. J., & Goicoecha, J. (2000). Sociocultural and constructivist theories of learning: ontology, not just epistemology. *Educational Psychologist, 35*(4), 227-241.

Padgett, D. K. (2012). *Qualitative and Mixed Methods in Public Health*. Los Angeles, CA: Sage Publications.

Pagano, R. R. (2010). Understanding Statistics in the Behavioural Sciences (8th ed.). Belmont, CA: Wadsworth.

Purdy, L., & Jones, R. (2011). Changing personas and evolving identities: the contestation and renegotiation of researcher roles in fieldwork. *Sport, Education and Society*, 1–19. DOI: 10.1080/13573322.2011.586688

Qu, S. Q., & Dumay, J. (2011). The qualitative research interview. *Qualitative Research in Accounting and Management, 8*(3), 238-264.

Richey, R. C., & Klein, J. D. (2007). *Collecting Data in Design and Development Research*. Mahwah, NJ: Lawrence Erlbaum Associates, Inc.

Ronglan, L. T., & Havang, O. (2011). Niklas Luhmann: coaching as communication. In R. Jones, P. Potrac, C. Cushion & L. T. Ronglan (Eds.), *The Sociology of Sports Coaching* (pp. 79–93). New York, NY: Routledge.

Schoenfeld, A. H. (1989). Ideas in the air: speculations on small group learning, environmental and cultural influences on cognition, and epistemology. *International Journal of Educational Research, 13*(1), 71-88. DOI: 10.1016/0883-0355(89)90017-7

Schonborn, R. (2000). What is technique? *Advanced Techniques for Competitive Tennis* (2nd ed.) (pp. 11–72). Oxford, UK: Meyer & Meyer Sport.

Silverman, D. (2001). *Interpreting Qualitative Data: Methods for Analysing Talk, Text and Interaction* (2nd ed.). London, UK: Sage Publications.

Taylor, J. B., James, N., & Mellalieu, S. D. (2005). *Notational Analysis Of Corner Kicks In English Premier League Soccer*. Paper presented at the science and football V: the proceedings of the fifth world congress on football, Lisbon, Portugal.

Trochim, W.M.K. (2006). Web Centre for Social Research Methods. Ithaca, NY. Cornell University. Retrieved July 20, 2015, from: http://www.socialresearchmethods.net/kb/index.php#about

Trumbull, M. (2005). Qualitative research methods. In G. R. Taylor (Ed.). *Integrating Quantitative and*

Qualitative Methods in Research (2nd ed.) (pp. 101–103). Lanham, MD: University Press of America Inc.

Trumbull, M., & Taylor, G. R. (2005a). Practical applications for developing research paradigms in quantitative and qualitative research. In G. R. Taylor (Ed.). *Integrating Quantitative and Qualitative Methods in Research* (2nd ed.) (pp. 222–229). Lanham, MD: University Press of America Inc.

Trumbull, M., & Taylor, G. R. (2005b). Major similarities and differences between two paradigms. In G. R. Taylor (Ed.). *Integrating Quantitative and Qualitative Methods in Research* (2nd ed.) (pp. 237–250). Lanham, MD: University Press of America Inc.

Turner, D. W. (2010). Qualitative interview design: a practical guide for novice investigators. *The Qualitative Report, 15*(3), 754-760.

Vasilachis de Gialdino, I. (2009). Ontological and epistemological foundations of qualitative research. *Forum Qualitative Sozialforschung / Forum: Qualitative Social Research, 10*(2). Retrieved from http://www.qualitativeresearch.net/index.php/fqs/article/view/1299/3163

Verstegen, M., & Marcello, B. (2001). Agility and coordination. In B. Foran (Ed.), *High-Performance Sports Conditioning: Modern Training for Ultimate Athletic Development* (pp. 139-166). Champaign, IL: Human Kinetics.

Viera, A.J., & Garrett, J.M. (2005). Understanding interobserver agreement: the kappa statistic. *Family Medicine, 37*(5), 360-363.

Webster, L., & Mertova, P. (2007). *Using Narrative Inquiry as a Research Method: An Introduction to Using Critical Event Narrative Analysis in Research on Learning and Teaching*. Oxon, UK: Routledge.